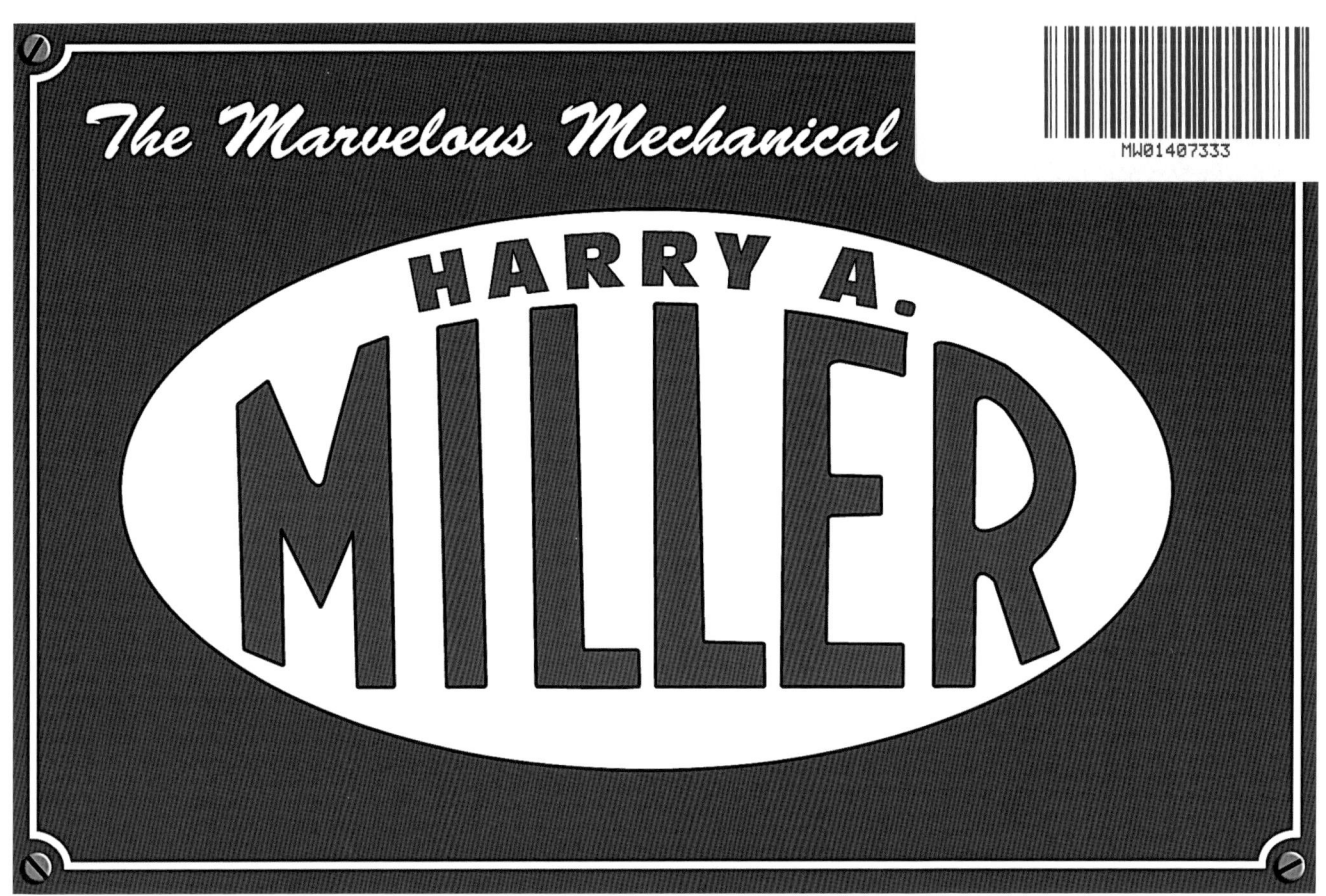

Gordon Eliot White

Iconografix

Iconografix
PO Box 446
Hudson, Wisconsin 54016 USA

© 2004 Gordon Eliot White

All rights reserved. No part of this work may be reproduced or used in any form by any means... graphic, electronic, or mechanical, including photocopying, recording, taping, or any other information storage and retrieval system... without written permission of the publisher.

The information in this book is true and complete to the best of our knowledge. All recommendations are made without any guarantee on the part of the author or Publisher, who also disclaim any liability incurred in connection with the use of this data or specific details.

We acknowledge that certain words, such as model names and designations, mentioned herein are the property of the trademark holder. We use them for purposes of identification only. This is not an official publication.

Iconografix books are offered at a discount when sold in quantity for promotional use. Businesses or organizations seeking details should write to the Marketing Department, Iconografix, at the above address.

Library of Congress Control Number: 2004102338

ISBN 1-58388-123-9

04 05 06 07 08 09 6 5 4 3 2 1

Printed in China

Cover and book design by Dan Perry

Copyediting by Jane Mausser

Cover photo-See p. 5

BOOK PROPOSALS

Iconografix is a publishing company specializing in books for transportation enthusiasts. We publish in a number of different areas, including Automobiles, Auto Racing, Buses, Construction Equipment, Emergency Equipment, Farming Equipment, Railroads & Trucks. The Iconografix imprint is constantly growing and expanding into new subject areas.

Authors, editors, and knowledgeable enthusiasts in the field of transportation history are invited to contact the Editorial Department at Iconografix, Inc., PO Box 446, Hudson, WI 54016.

TABLE OF CONTENTS

	ACKNOWLEDGMENTS	4
Chapter 1	HARRY MILLER: AN ARTIST IN METAL	6
Chapter 2	RACING CARS	18
Chapter 3	RACING ENGINES	32
Chapter 4	RACING BOATS	42
Chapter 5	PASSENGER CARS	58
Chapter 6	LAND SPEED RECORD CARS	69
Chapter 7	LATER RACING ENGINES, 1930–1933	78
Chapter 8	EXPERIMENTS THAT FAILED	88
Chapter 9	MILLER-FORDS	101
Chapter 10	THE GULF MILLERS, THE "CARS FROM MARS"	108
Chapter 11	LSR, L-510, CHRISTIE COMBAT CAR, MILLER MIDGET	118
	EPILOGUE	126

ACKNOWLEDGMENTS

In doing the research for my four earlier books, Offenhauser, The Indianapolis Cars of Frank Kurtis, KURTIS-KRAFT, and Lost Race Tracks, I had the help of literally scores of individuals going back over many years. This book is different. Although I acknowledge the advice and education I have gotten over the years from a great many people, this book owes its existence to a mere handful. Chuck Davis, Jim Etter, Dave Uihlein, Buck Boudeman, and Dean and Don Butler were of great help, but two people made this book possible. The late Mark L. Dees and the late Griffith Borgeson preserved the memory of Harry A. Miller when he was all but forgotten. Mark's heirs bequeathed to me his collection of Miller photographs and they and the late Bob Sutherland made me the archivist of the surviving Miller drawings, most done by that incomparable engineer and draftsman, Leo W. Goossen.

I added to my knowledge of things Miller by the research for my 1996 Offenhauser book and, over the intervening eight years, other material has come to light that I have incorporated here. Thanks too, to Ron McQueeney at the Indianapolis Motor Speedway for permission to use old photographs to which the Speedway holds the copyright. I want also to recognize Harold Peters who is helping carry on the Miller tradition through a web site, http://www.milleroffy.com. All of the original drawings reproduced here are the property of the Leo Goossen Archive, of which I am the curator. Because of their age, some of these drawings have become faded and stained, but nonetheless I believe they are worthy of being published.

Both Mark's and Griff's Miller books are now out of print, although Griff's Golden Age of the American Racing Car has been re-issued. This book cannot replace either Mark's tour de force *Miller Dynasty* nor Griff's much smaller *Miller*, but I hope that it will bring the story of Harry Miller, his high art and his broad influence on powerful American machinery, to the attention of many readers who may not have known of them.

Harry Armenius Miller in 1927 at the height of his powers, with a supercharger for one of his 91-ci engines. The 8-inch impeller turned 7,000 rpm.

Chapter 1: HARRY MILLER: AN ARTIST IN METAL

Harry Arminius Miller was a self-taught mechanic from Menomonie, Wisconsin, who learned his trade in the shops that served the lumber camps. At that time, in the 1890s, Americans worked largely by trial and error to "see what works." With determination and a level of mechanical genius, they invented the modern world. Scores of tinkerers and inventors such as Edison, Ford, and the Wright Brothers combined common sense and native ingenuity to build the devices that turned the new twentieth century into the American Century. In Europe, by contrast, a far more elevated mode of invention involved academically trained engineers. Miller, son of a German immigrant, was one of the tinkerers.

Born in 1875, he left home as a teenager and worked in Minnesota, Utah, Idaho, and Los Angeles. He married Edna Lewis in 1897 and returned to Menomonie, where he was hired by the Globe Iron Works and rose to foreman of the foundry.

Miller next went to Toledo and was employed by a short-lived company that built the Yale automobile, then moved to Lansing, Michigan, where he became a mechanic on Ransom E. Olds' Vanderbilt Cup racing team. Olds did poorly in the 1906 Cup races and Harry returned to California where he opened a small machine shop that specialized in carburetors. He developed and patented several designs for the combining of air and gasoline into proper mixtures to fuel the new internal combustion engine.

Branching out into the manufacture of automobile pistons and fuel pumps, cast from an aluminum alloy of his own devising, Miller's shop took in job work from the automobile racers in auto-mad Southern California. His carburetors were successful among the racers and became common on winning cars. Barney Oldfield and Bob Burman were among those who also turned to Miller to make major engine repairs after blowups on the track.

In 1913, a talented young machinist from the Pacific Electric Railway shops, Fred Offenhauser, applied for a job at the Harry A. Miller Mfg. Co. Offenhauser would prove to be not only a superb machinist, but also a balance wheel to Miller's inventive imagination for the next two decades.

A year later, Eddie Rickenbacker blew up the engine in his Peugeot, then the world's best racing car, and unloaded the remains on Miller. Shortly afterward, Burman's Peugeot also broke a connecting rod. Miller and Offenhauser undertook to rebuild both engines. The new Miller-Peugeot proved remarkably fast and led Oldfield to ask Miller to build not only an engine, but also an entire racing car to a fantastic, aerodynamic design that would become known as the "Golden Submarine." Although the Sub was only moderately successful, it was new and spectacular and other racers flocked to Miller's shop for engines, parts, and repairs until the World War came to the United States on April 6, 1917 and interrupted auto racing for the duration.

When the United States entered World War I, virtually every establishment that could build things mechanical turned to war work. At that time, no U.S.-built aircraft engine existed or was under development with enough power to operate a pursuit plane or even an advanced trainer. American industry undertook to build what became known as the Liberty engine, a 330-horsepower V-12. In the interim, the U.S. Army attempted to get licenses from the Allies to make the Rolls-Royce, Hispano-Suiza ("Hisso"), and Gnome rotary engines. The difficulties, both of making the commercial arrangements and in converting European drawings and specifications to U.S. practice, were immense. The result was that no Rolls engines were built here during World War I and only relatively small numbers of Hispano-Suiza and Gnome engines were manufactured in the United States, all for training aircraft.

Assistant Secretary of War Benedict Crowell reported that when an Army mission went to France to look at aero engines it was shown an apparently advanced U-16 design by the Bugatti factory said to develop the then very respectable output of 510 horsepower. The engine consisted of two eight-cylinder units with separate crankshafts geared together in a common crankcase. The engine was tested for the Army and although an American soldier walked into the spinning propeller and was killed, the United States agreed to buy the design. The accident bent one of the engine's two crankshafts; the damage was not found until it reached the United States.

The engine, specifications, and drawings were quickly shipped to New York City and Duesenberg, Miller, and other companies were hired to adapt the design for American manufacture. Because of certain peculiarities of the engine and the discovery after it reached the United States that much development-work had yet to be done, a major redesign was undertaken by engineer Charles B. King. August and Fred Duesenberg subcontracted the carburetors and fuel pumps to Miller, who, with Offenhauser, moved to New York City to carry out manufacture of the parts. Working out of quarters at 109 West 64th Street (now part of the Lincoln Center complex) Miller was in residence in New York City until February 1919, a little less than eight months. Despite the redesign and a great deal of work by the Duesenberg brothers at their Elizabeth, New Jersey, plant, the Bugatti U-16 never ran reliably. The crankcase gears were inadequate and torsional vibrations would strip the gears, which allowed the crankshafts to collide, which in turn destroyed the engine. Only 11 engines, of dubious quality, were produced by the time of the Armistice, November 11, 1918.

When the war ended, Miller returned to Los Angeles, poised to become the premier builder of racing cars and engines in the United States.

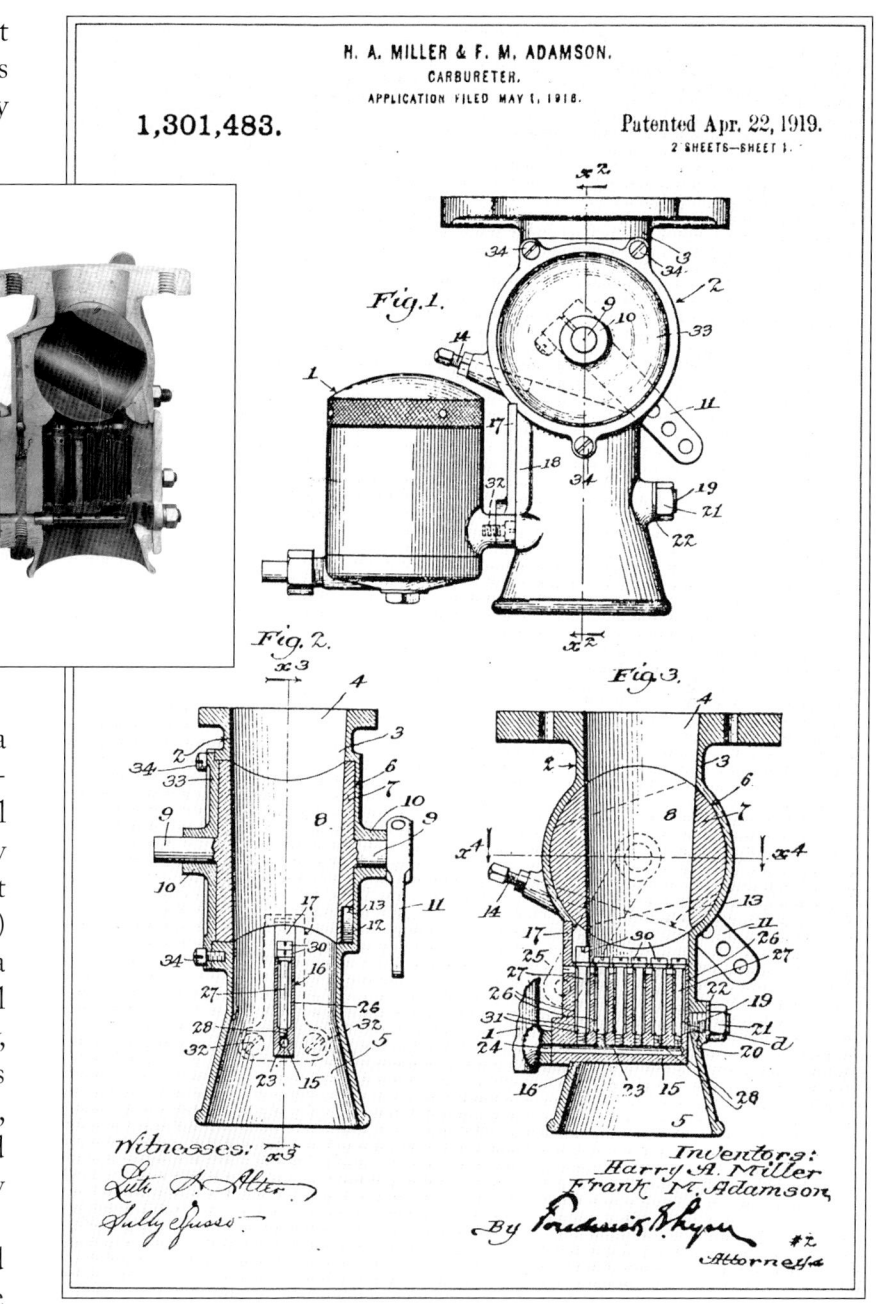

A Miller type "H" carburetor of 1926, and a patent drawing of the 1919 version.

Between 1914 and 1920, fuel pumps for the better engines were complex, expensive devices. This is similar to the design of the pump that Miller and Fred Offenhauser manufactured in New York City in 1918 for the Bugatti U-16 engine that the Duesenbergs were building under an Army contract.

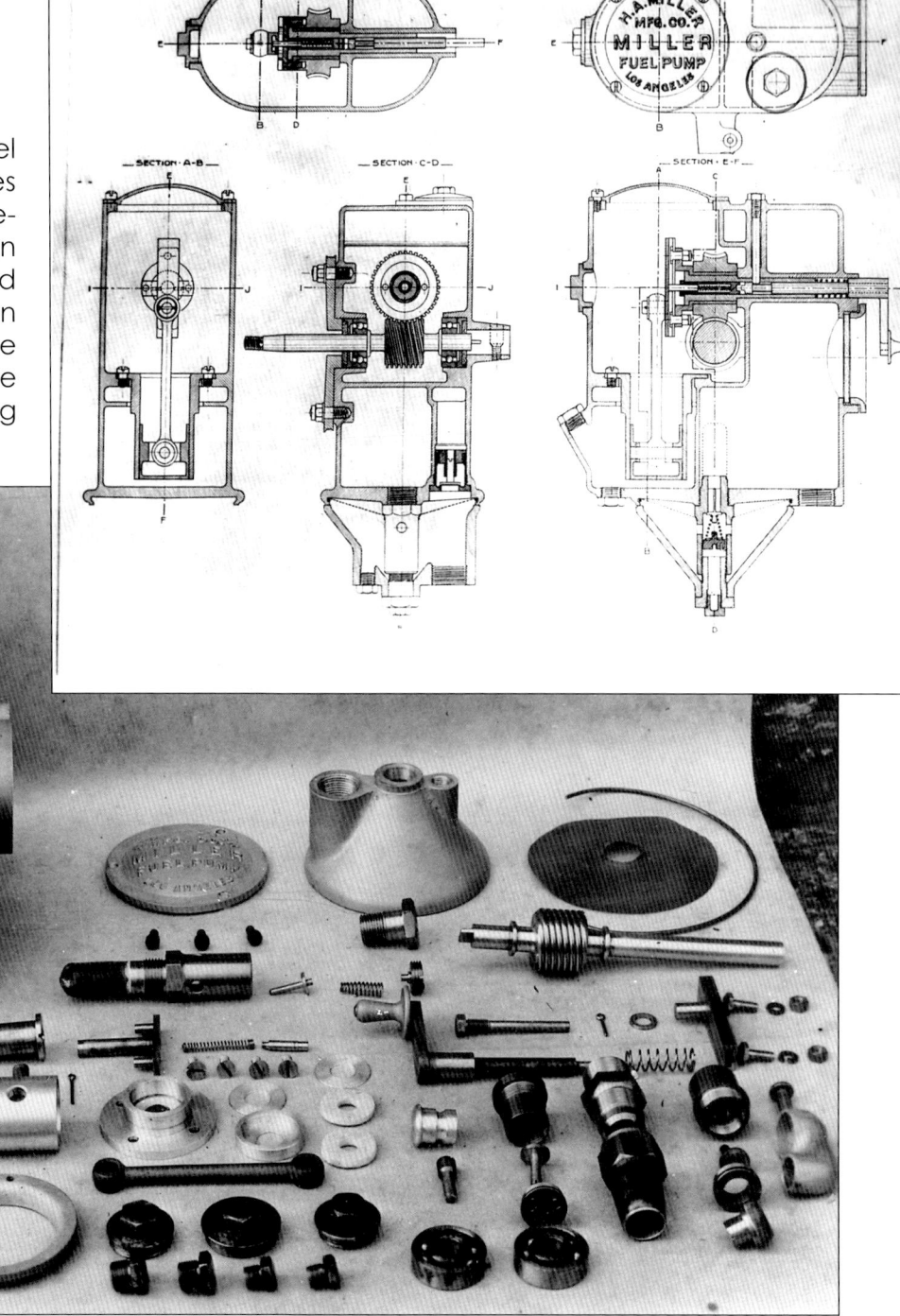

This, believe it or not, was Miller's first dynamometer, a portable device with a water turbine to absorb power, a water supply tank, and a scale to measure torque. The engine under test is a single-cam four, the "Golden Submarine" engine.

Miller's plant on Long Beach Avenue in the 1920s was not modern. Even so, it turned out some of the most advanced engine designs of its time on heavy lathes such as the big Monarch in the foreground and a Van Norman vertical mill capable of handling hefty engine blocks. Here Fred Offenhauser and Ernie Olson look over the crankcase for the W-24 marine engine.

When Bob Burman, the speed king of the day, blew up his Peugeot in Los Angeles in 1914, he turned to Miller rebuild and improve it. The result was the first of a long line of highly innovative and successful Miller racing engines.

The 1915 "Golden Submarine," built for Barney Oldfield, was fantastically advanced for its day: aerodynamic with an enclosed body. Barney was most successful with in match races against Ralph DePalma in his Packard.

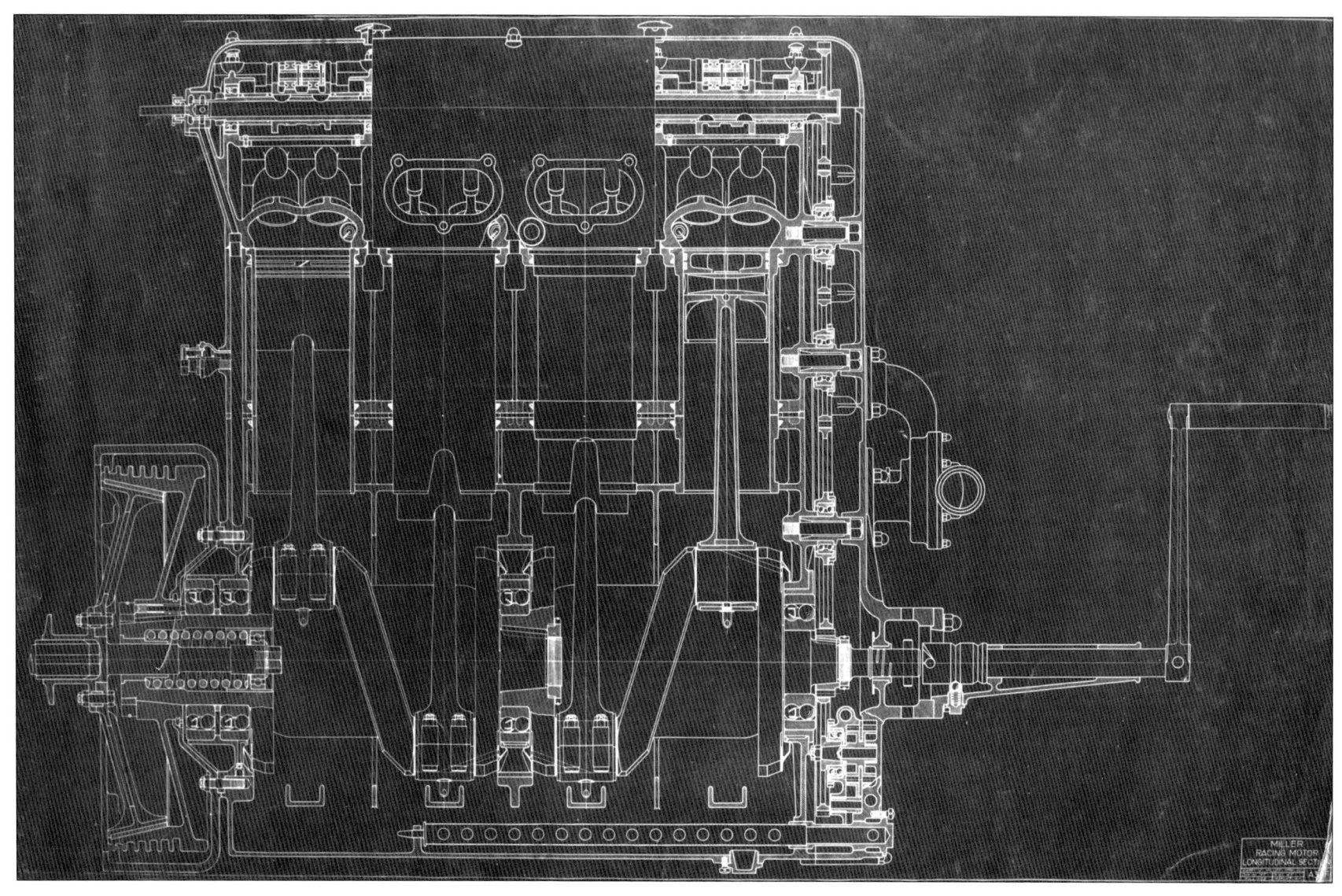

The Sub was powered by a 289-ci single-cam four, seen here in a cross-sectional drawing.

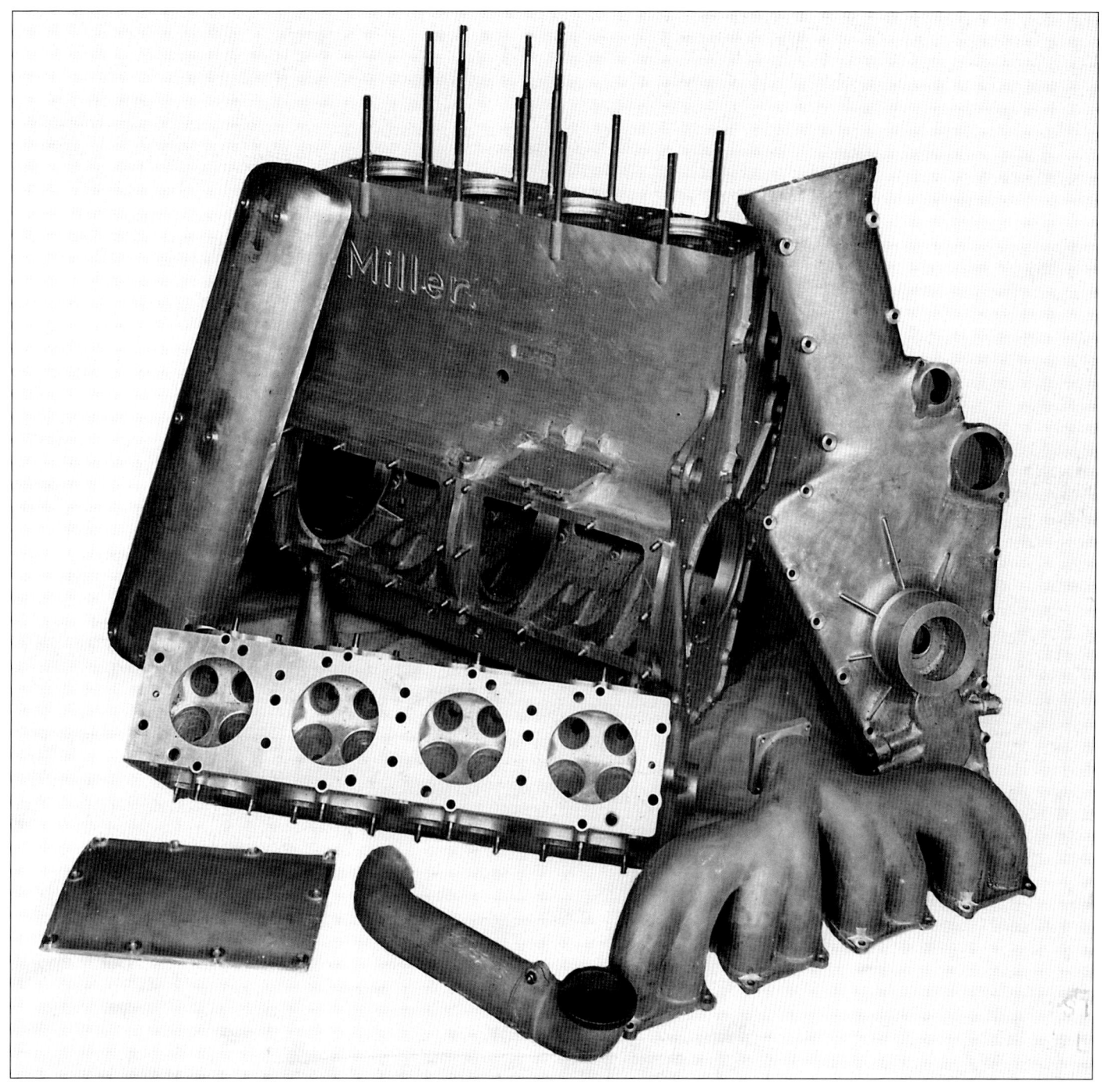

A disassembled view of the Sub's engine. Although it used a single overhead camshaft, its four-valve pent-roof combustion chamber was quite modern. A similar engine was built for aircraft use with a propeller boss instead of a clutch.

When Miller and Offenhauser moved east to help in the armament effort for World War I, they were put to work in New York City building carburetors and fuel pumps for the Bugatti U-16 aero engine, one of the world's worst engine designs.

Leo Goossen, seen here standing at the back of Miller's drafting loft at the Long Beach Boulevard plant, drew the designs for almost all of Harry Miller's wondrous creations in this dark, drafty room.

Chapter 2: RACING CARS

The vehicles that made Miller's reputation in the 1920s were his Indianapolis racing cars. The 183, 122, and finally the beautiful 1926–1929 91s burst upon the racing scene like a skyrocket.

The lines of his racing car's bodies moved American racing vehicles from the square, upright era of the 1910s into a low rakish look that precisely fit the era, a time of bootleg gin and Gatsby-like luxury. The fit and finish of Miller's cars was every bit as good as his engines. Look, for example at the lineup at Indianapolis in 1922, at the Miller-built cars of Frank Elliott and Cliff Durant. Compare them to the cars of Art Klein, Eddie Hearne, and Leon Duray for beauty and line. There is no comparison.

In 1919, Cliff, son of General Motors founder Billy Durant, paid Miller $27,000 to build a new racing car, for which Miller designed a single-cam four of 182 ci, only slightly updated from the "Golden Submarine's" engine of three years earlier. This engine, built before Leo Goossen appeared on Miller's doorstep, was a failure—the last failure Miller would have for more than a decade. The car, known as the "Baby Chevrolet" to advertise Durant's then-current corporation, was beautifully finished but slow.

Goossen's first assignment when he went to work for Miller was engines for two racing cars for Edward Maier, a Los Angeles brewer. Maier's racing operation was known as the TNT Auto Co. and the engines were known by that title. They were the first twin-overhead-cam Miller engine design, using flat-topped valve cups. Neither T.N.T. made it to Indianapolis, but one of the chassis survives. Miller subsequently built two somewhat similar cars for Riley Brett, known as Junior Specials, in which Brett mounted twin-cam six-cylinder engines of his own design.

Indianapolis notified the racing community in mid-1922 that the 1923 formula would allow only 122 ci displacement "in order to cut down on the terrific speeds" that were being turned—Murphy had qualified at over 100 mph! Miller, Goossen, and Offenhauser accordingly got to work and produced a 121.5-ci straight-eight similar to the 183.

The last of the Miller 183s was delivered to Tommy Milton early in 1923. He used it to set a record of 151 mph at Muroc, California, on April 4, 1924.

The chassis, into which the new 122 engines were installed, were one-man cars, extremely narrow (just over 21 inches wide), and had a frontal area of only 5 square feet. The Miller 122 was the first racing car to be serially (one hesitates to say "mass") produced, with seven built before the Indianapolis race of 1923. Myron Stevens, a true artist in aluminum, was chief of Miller's body shop and the design set the style for American racing cars for a generation. Small parts were cast, machined, and polished as though they were fine jewelry.

The Duesenberg brothers brought four supercharged cars to the Speedway in 1924 and one of their drivers, Joe Boyer, driving in relief of L. L. Corum, won the race. Miller immediately went to work on his own supercharged engine. Although the blown Miller 122 did not win the 1925 race, Millers took six of the first nine places that year. The even more "terrific" speeds, now 113 mph, led to the cut in displacement to 91 ci (1-1/2 liters) for 1926.

The search for ever more power out of ever-smaller engines having been essentially solved, Miller turned to the chassis in a search for more speed through better handling. The result was front wheel drive. Unlike the cross-mounted engine of the front-drive Christie that Barney Oldfield had used to blast around a hundred dirt tracks, the Miller's engine was mounted on the center line and the power transmitted via a transaxle that included the transmission and differential.

Removing the driveshaft from the cockpit allowed the Miller drivers to sit 9 inches lower than in the rear-drive cars, with a corresponding reduction in frontal area. The resulting low body with its incredibly long hood would attract automotive stylists for a generation. E. L. Cord immediately, and, in Europe, Citroen a decade later, would adopt front wheel drive. The automotive world at large came to front drive in the 1970s, after the Arab oil embargo raised gasoline prices to the point that weight saving and aerodynamics became a necessity, but Miller was the first to apply it successfully, as the Smithsonian Institution's 1929 Leon Duray Packard Cable Miller Special will attest.

The front-drives were the last of the classic Miller racing cars. They were outmoded by the change in the Indianapolis rules for the 1930 race. Miller sold his business to Schofield and when he returned to racing his world had been turned upside down and severely shaken by the Depression. To meet the new situation he designed a new larger engine, the "Big Eight" of 230 ci, and a larger two-man chassis to carry it. Those cars were more massive than the lithe rear drives of the 1920s and significantly taller than the front drive cars. Because economics did not permit the luxury of the costly front-drive mechanism, most were rear-drives and, of course, they were not supercharged. To save on cost the engines had two valves per cylinder.

Several variants were built in the early 1930s including a 303-ci V-16, now nicely restored by Chuck Davis and 308-ci, four-wheel drive car now in the possession of Dean Butler. The second four-wheel drive machine was not sold until Miller had sunk into bankruptcy and it was knocked down to $1,600 for Frank Scully. After competing in Europe, it was wrecked and the engine ended up with Bunny Phillips who installed it in a type 35 Bugatti that he ran at Indianapolis before World War II.

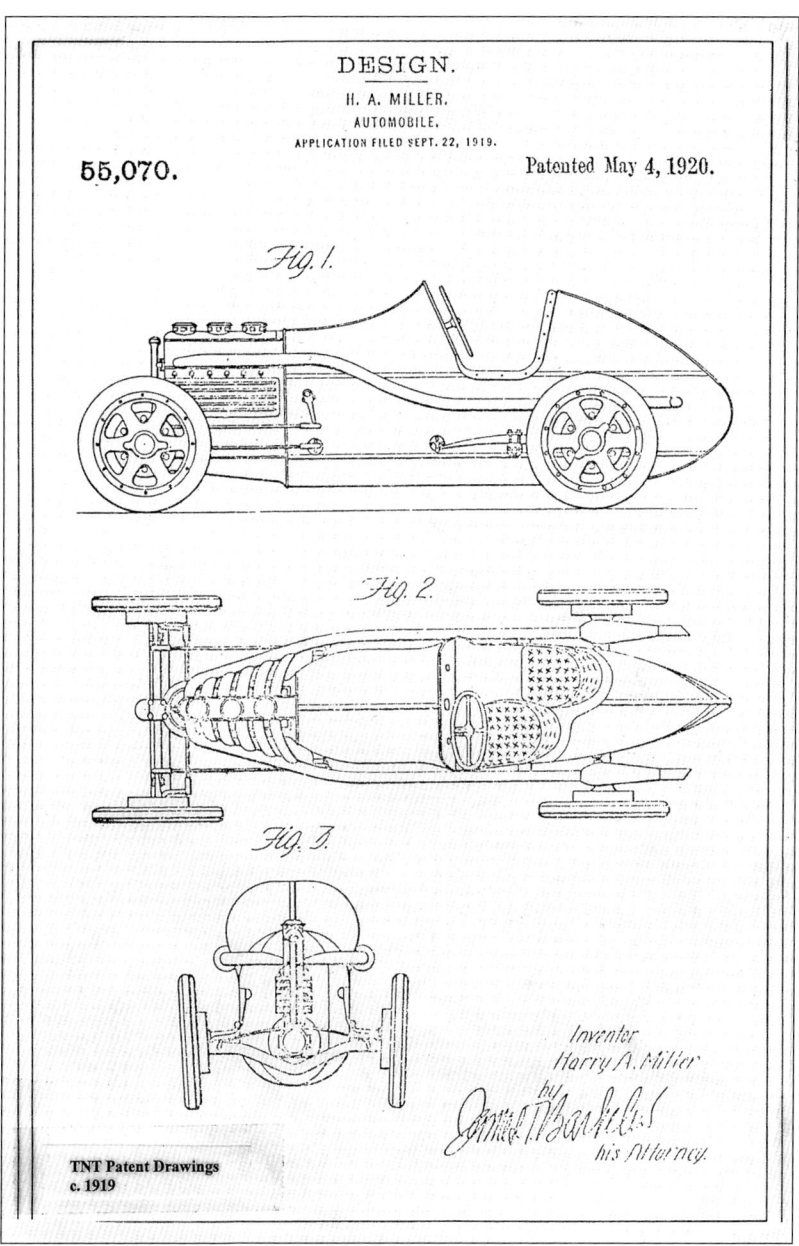

Miller filed this patent drawing in 1919 for the T.N.T. car. The fanciful front end is the artist's conception, but the rest of the body reasonably resembles the car as Miller built it.

The T.N.T. chassis was an extension of Miller's fixation on aluminum castings. The body was built of cast sections, although the hood, in the end, was sheet metal. The car was tested at the Beverly Hills board track and driven in a few races by Frank Elliott. While the T.N.T. engines have disappeared, the chassis has survived, now equipped with a Miller 183 engine.

Following on the modest success of Oldfield's "Golden Submarine," but without the unconventional closed body, Miller built this similar car for A. A. Cadwell, shown here in 1916 with Reeves Dutton, Miller, and Claude French, a Miller employee. Cadwell campaigned it indifferently in 1917 and 1918, his efforts being hampered by the AAA's suspension of racing because of World War I. Omar Toft bought it and ran it in 1919, after which it went to J. Alex Sloan and was destroyed along with the Submarine when Sloan's barn in Joliet, Illinois, burned to the ground.

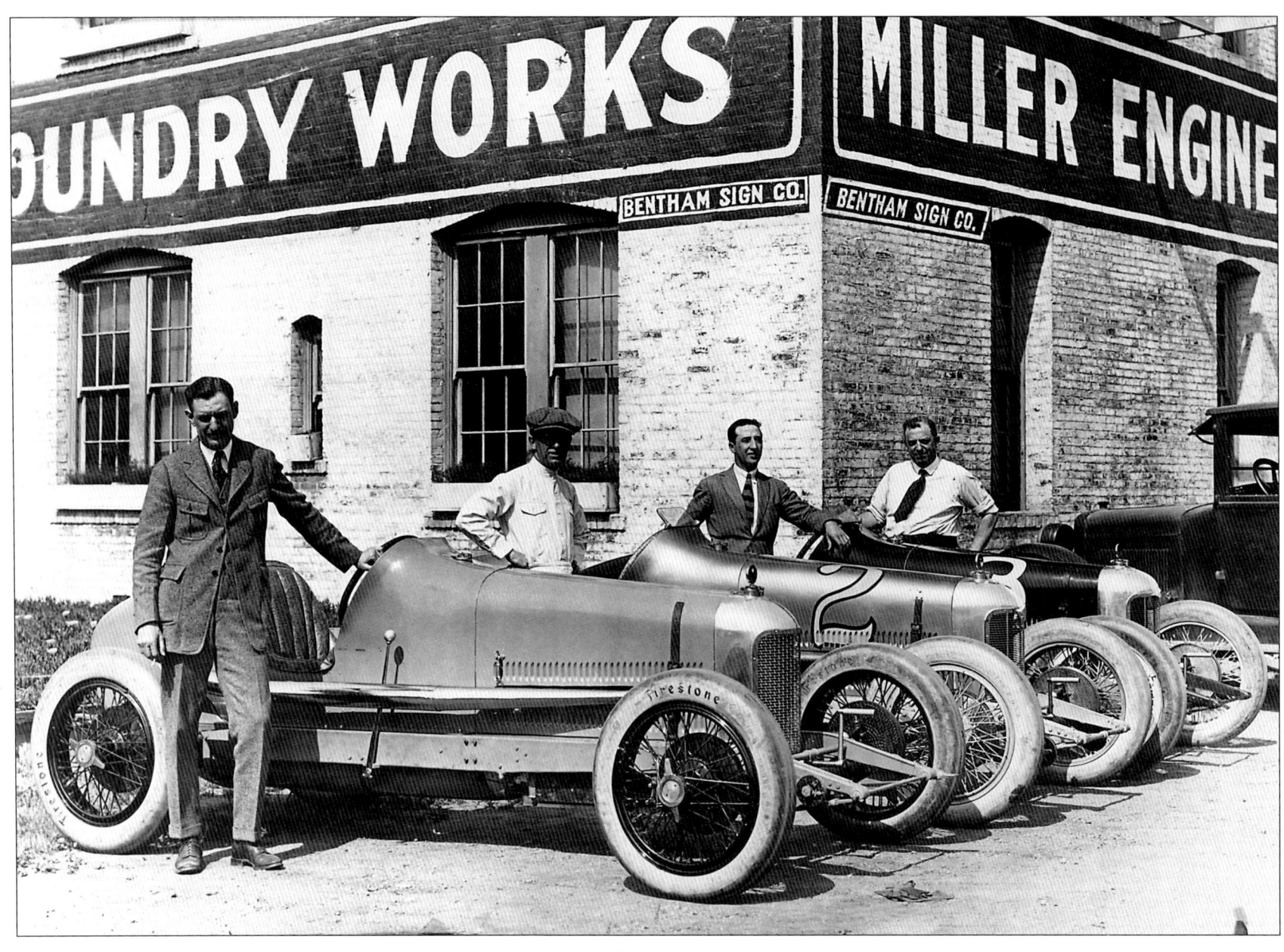

A bevy of Miller 122 rear-drive cars seen outside Miller's Long Beach Avenue plant in 1924. Left to right are Ira Vail, Jimmy Murphy, Martin D. Alzaga, and Miller.

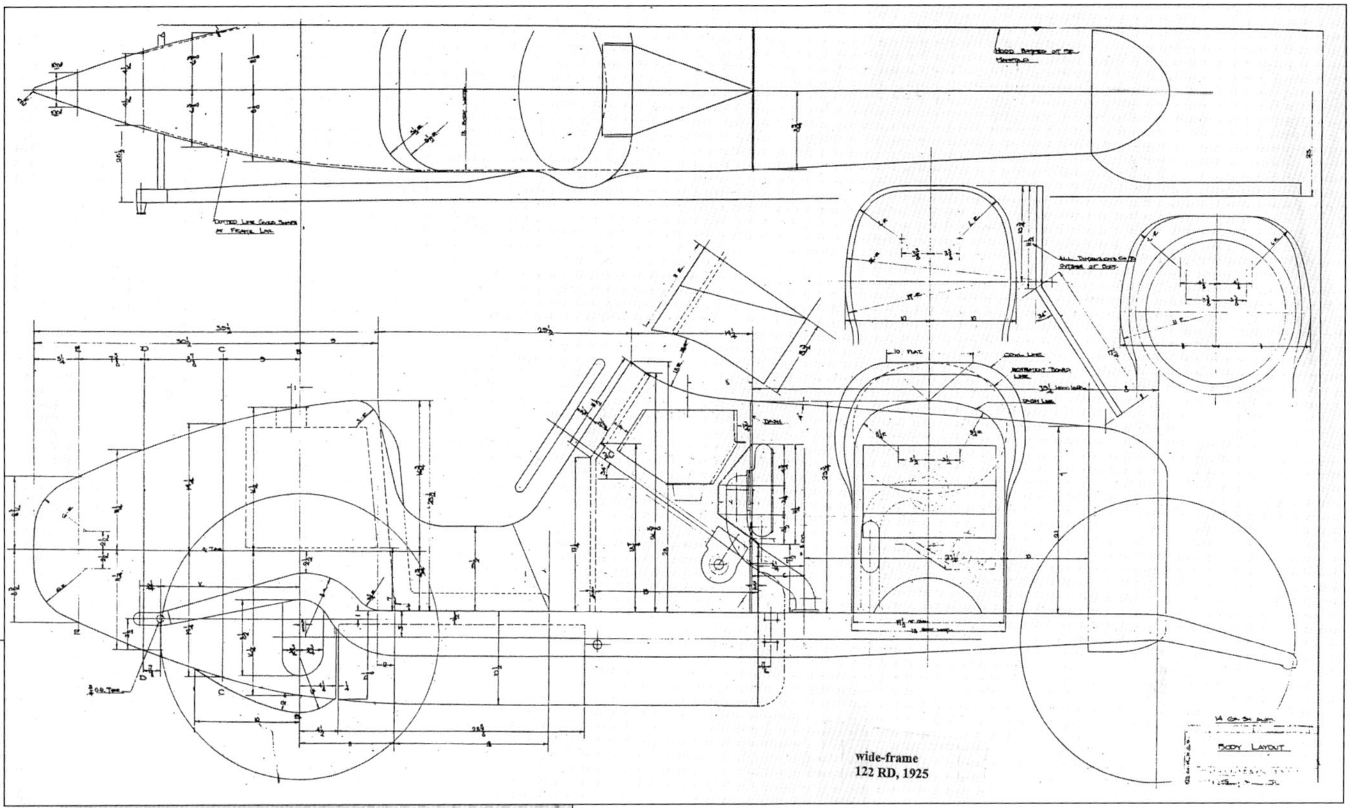

wide-frame
122 RD, 1925

Harry Hartz's 122 Miller chassis at Indianapolis in 1924, when he finished fourth. Although entered by Cliff Durant, it was entirely built by Miller's craftsmen. The drawing shows the Miller body and frame layout.

Frank Lockhart's 1927 91 Miller rear drive car. Lou Meyer was driving this car when he finished second at Indianapolis in 1929.

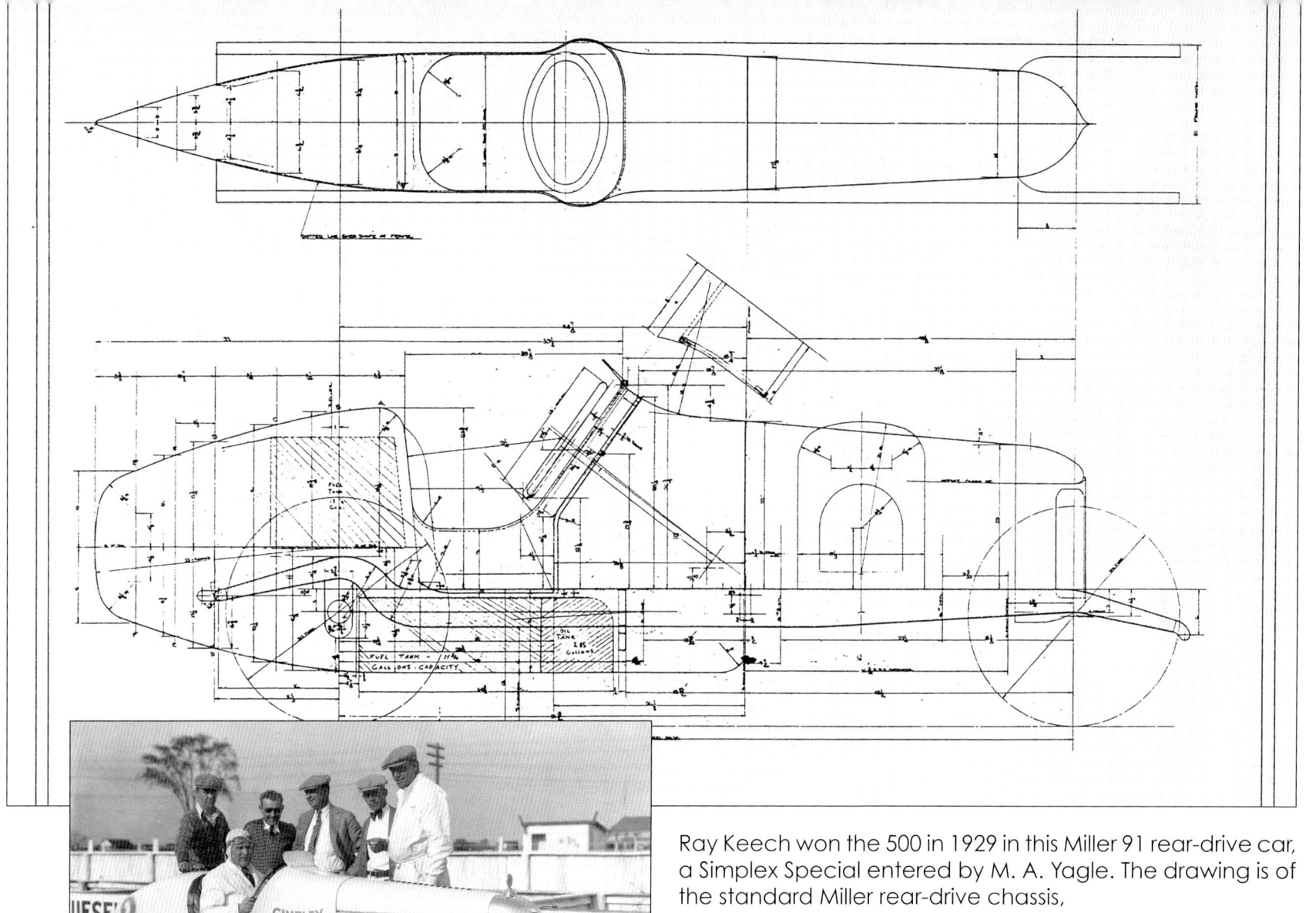

Ray Keech won the 500 in 1929 in this Miller 91 rear-drive car, a Simplex Special entered by M. A. Yagle. The drawing is of the standard Miller rear-drive chassis,

This was Cliff Woodbury's rear-drive Miller 91, a Boyle-sponsored entry.

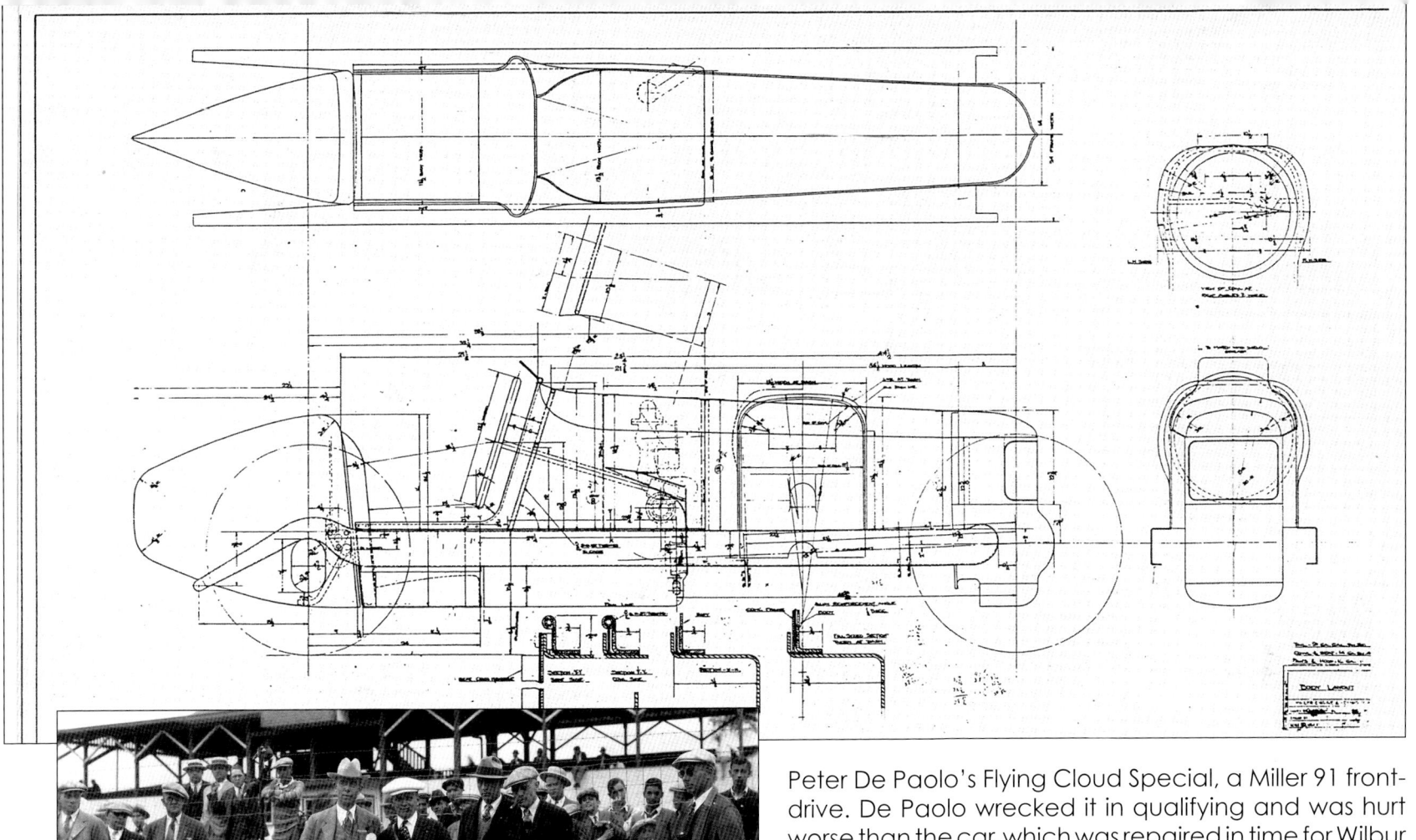

Peter De Paolo's Flying Cloud Special, a Miller 91 front-drive. De Paolo wrecked it in qualifying and was hurt worse than the car, which was repaired in time for Wilbur Shaw to drive it to 25th place when the engine stripped its timing gears.

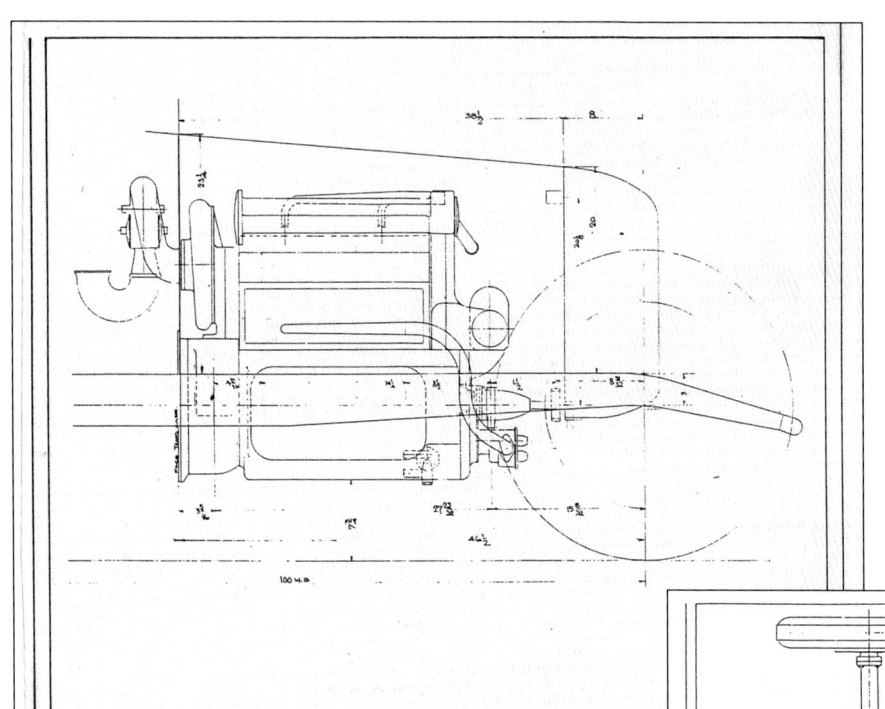

Drawing for a 151 Miller Marine engine mounted in a 122 chassis to run on the dirt tracks.

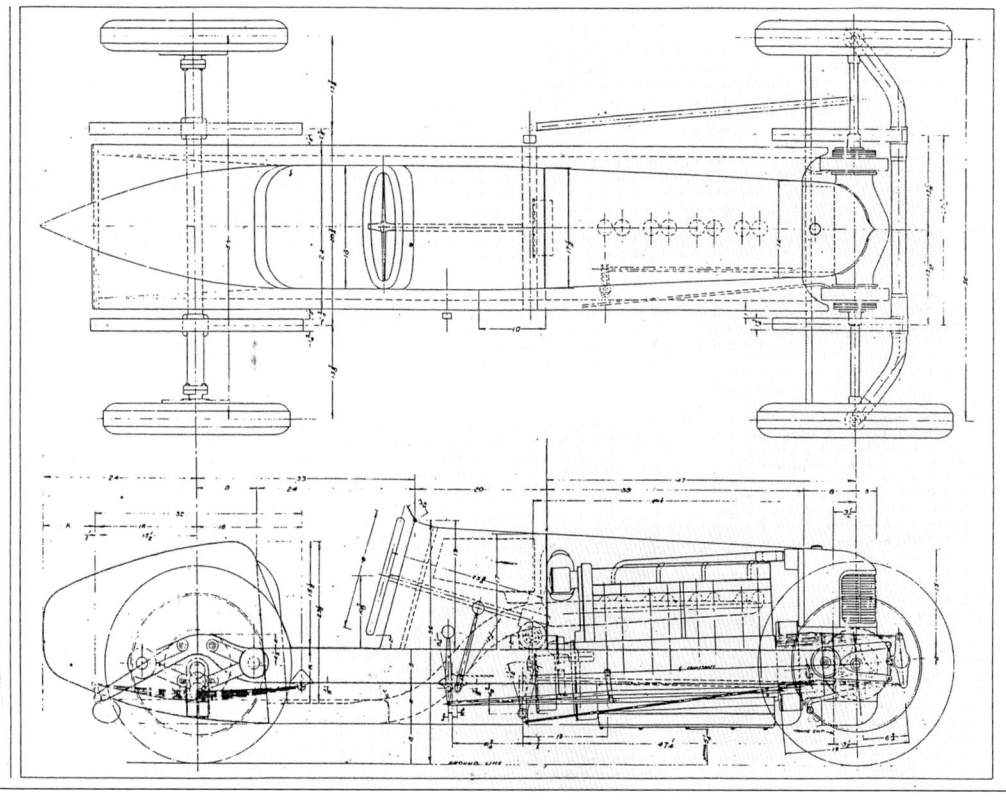

The original prototype drawing, done in 1924, for the 122 Miller front-drive. It was substantially modified before the first cars were built.

Dave Lewis in one of the front drive Miller "Junior Eight" cars of 1925.

Chapter 3: RACING ENGINES

Harry Miller conceived, designed, and built at least 37 different engines along with supercharged, unsupercharged, and marine variants. He may well have built more. He had others designed and drawn up by Leo Goossen that were never built. All told, Miller may have manufactured 200 to 225 examples over a 22-year career. For a small plant, indeed not much more than a shop for most of Miller's career, his was an impressive output of ideas and metal.

The first engine of Miller's own design, although based in part on the Burman Peugeot of 1915, was that of the "Golden Submarine" described earlier. The Christofferson brothers went to Miller's shop to have built a six-cylinder engine for their experimental airplanes. Two years later, in April 1917, Miller himself designed and built a 1,414-ci V-12 aimed at the military aviation market. The Army tested the prototype but turned Miller down on its production.

The most notable event in the history of the Miller design team was the hiring of Leo Goossen as a draftsman and engineer in July of 1919. Goossen had worked for Walter Marr at Buick and had an excellent idea of what was practical and what was not in automotive design. By the time Goossen arrived the always-practical Fred Offenhauser had become boss of the machine shop. From the end of 1919 through the summer of 1933, when Miller's company failed and Leo left his employ, a particular spare architecture was seen in Miller's engines. Gone were the earlier cumbersome long-stroke models and the more bizarre designs such as the desmodromic roller cam of the aero V-12. After Miller left California, his higher flights of automotive fancy reasserted themselves at Miller-Hibbard, Gulf, and the Tucker Aircraft Co.—none of them successful.

The first real gem in the Miller crown was the 183-ci eight-cylinder racing engine. In 1920, Tommy Milton, who would win the Indianapolis 500 a year later, used Miller's Los Angeles shop as his base. Unhappy with the Baby Chevrolet and even more unhappy with the Duesenberg brothers in a dispute over a land speed record car he bought from them, Milton persuaded Miller to design a straight-eight engine that borrowed from both Duesenberg and Ballot practice. What they built was a twin-overhead-cam straight-eight with four valves per cylinder. It was arranged with two four-cylinder blocks atop an aluminum crankcase. The cams were driven by a train of gears that also operated the water pump and magneto. This was the same basic arrangement that prevailed for Miller's engines until the early 1930s and for the entire history of the Offenhauser engines.

Initially the 183 engine's cylinder heads were detachable, but later, as with the engines that followed, the heads were cast integrally with the block. That basic engine architecture set the scene for all of the Miller and Offenhauser engines for the next 55 years. With Miller's inspiration, Goossen's engineering and drafting skill, and Milton's racing experience, the 183 was immediately successful. The first year the Miller engine was available, Jimmy Murphy mounted it in a Duesenberg chassis and won with it at Indianapolis in 1922.

Another early Miller design was a special head for a hot Model T Ford–powered car then being run on local dirt tracks by Col. Harry Hooker. The T featured full pressure oiling, achieved ingeniously at first by converting one of the unused flathead valve lifters into a pump. Another version used gears to drive the cams while a third version used a less expensive timing chain.

The 183-ci-engine rule at Indianapolis changed for 1923, with engines reduced to 2 liters (122.042 ci). Miller thus designed a smaller engine, based largely on the 183 although lighter in weight with only two valves per cylinder and five main bearings. The 122 won at Indianapolis in 1923 and even though Duesenbergs won the 500 in 1924 and 1925 with supercharging, Miller 122s took six of the top ten places in 1924 and six in 1925.

The Indianapolis rules reduced engine displacement to 91 ci for the 1926 season but the supercharged Miller 91 produced even more power than had the 122. The supercharged 91 was the pinnacle of Miller's art. On the dynamometer, it produced the amazing output of 1.7 horsepower per cubic inch and weighed but 2 pounds per horsepower, a very advanced engine by the standards of 1926. A V-16 of 91 ci was designed in 1926, but the project never found a sponsor.

Both the 122s and the 91s were sold in marine versions for race boats. Miller developed a 310-ci eight for boat racing in larger classes and the 310 powered Arron de Roys' hydroplane Lady Helen II to victories in the Gold Cup and President's Cup races in 1926.

One of the four-cylinder blocks from the 310 was used to make a 151-ci four-cylinder Miller marine engine for the smaller-displacement classes. The 151 marine engine was quite successful in boat racing circles and was quickly adopted by dirt track racers. It formed the basis for the later Miller 200, 220, and 255 fours and was thus the ancestor of the Offenhauser 255- and 270-ci engines.

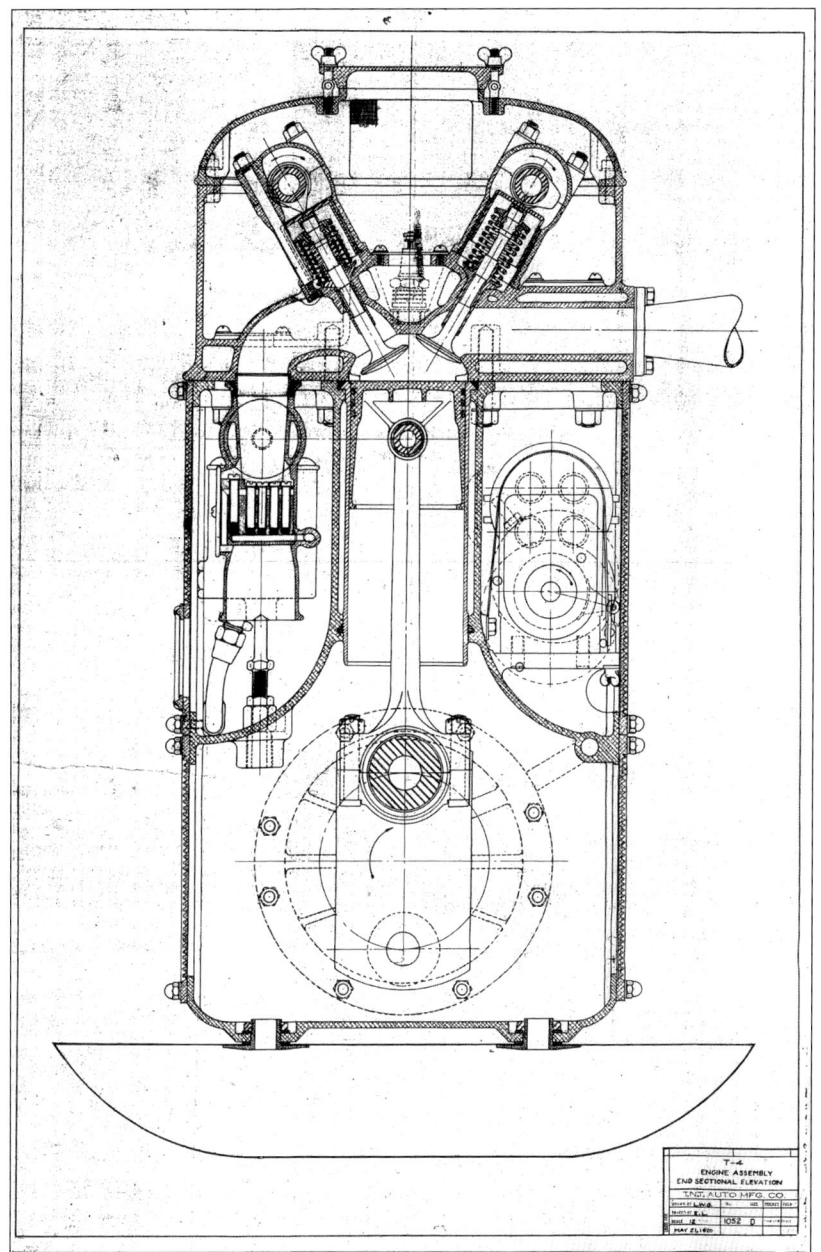

The 1920 T-4 engine, built for the TNT Auto Mfg. Co., Harry Miller's customer in Los Angeles. The 166-ci T-4 was a huge aluminum casting, with a complete enclosure of the entire engine. It displayed the basic Miller architecture, although Leo Goossen and Fred Offenhauser would shortly disabuse Harry of the enclosure and make a myriad of other necessary modifications to the basic Miller engine architecture.

Col. Harry Hooker had Miller build three different overhead cam heads for the Ford Model T engines in his several No. 99 cars. The one shown here used a chain to drive the cams. From Ford T frames to the remains of a Miller 122 Indianapolis car, Hooker did very well with cheap and cast-off parts. The Ford crankshaft, however, was not up to the power Miller's modifications produced and Hooker had to have a special crank hewn out of a steel billet. (The Hooker 122 frame has been retrieved in recent years and a complete, magnificent Miller 122 reconstructed upon it.)

The first 183-ci Miller engine was built to driver Tommy Milton's order in 1920. It was Harry's first eight, which, with more piston area, had inherent advantages over a four, or a six. Miller took the valve train design from contemporary Ballot practice, the cam design from Hall-Scott, and the lower end from Duesenberg. Ira Vail bought the second 183. Both of those engines had detachable cylinder heads with four valves per cylinder. Later 183s had their blocks and heads cast as one. Jimmy Murphy won with this engine at Indianapolis in 1922. The 183 generally used a Delco distributor driven off a gear in the center of the exhaust cam. With battery ignition, cars needed a generator, which was gear-driven from a power take-off on the front of the engine. The carburetors on this engine are Miller's Master brand.

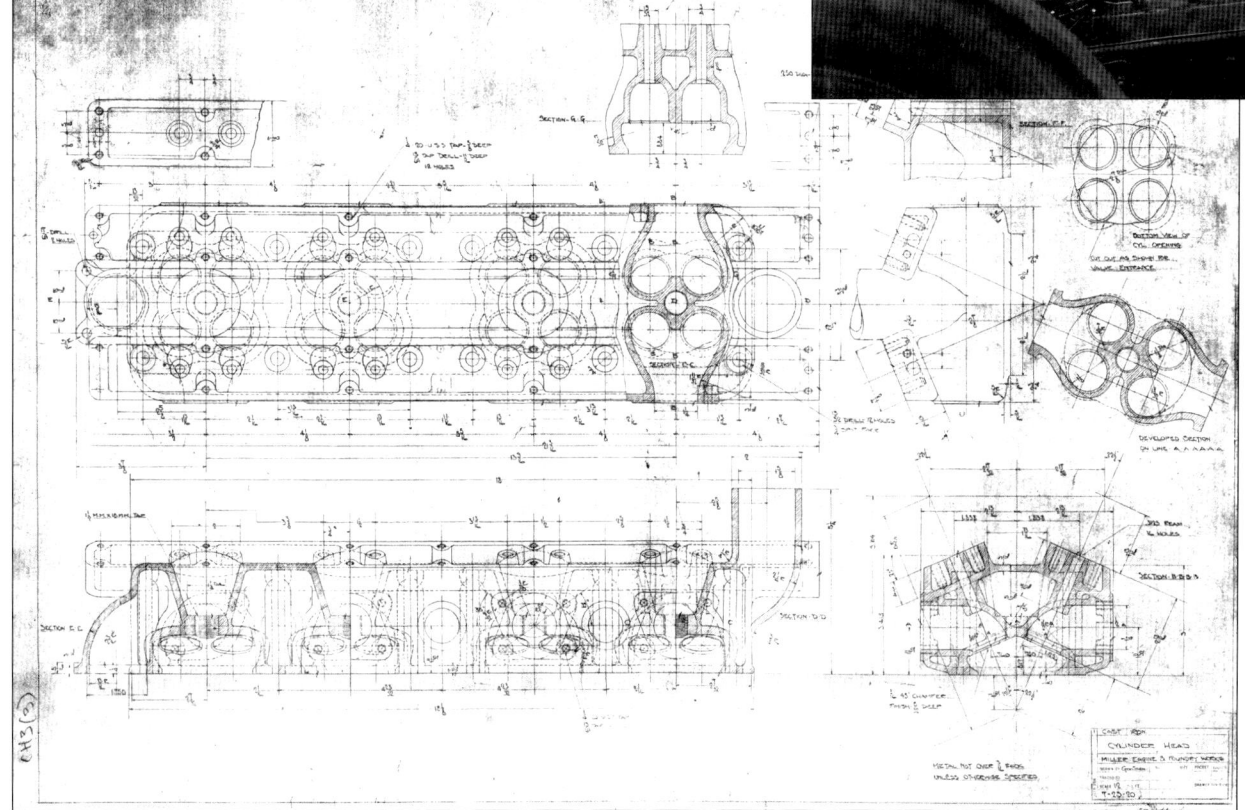

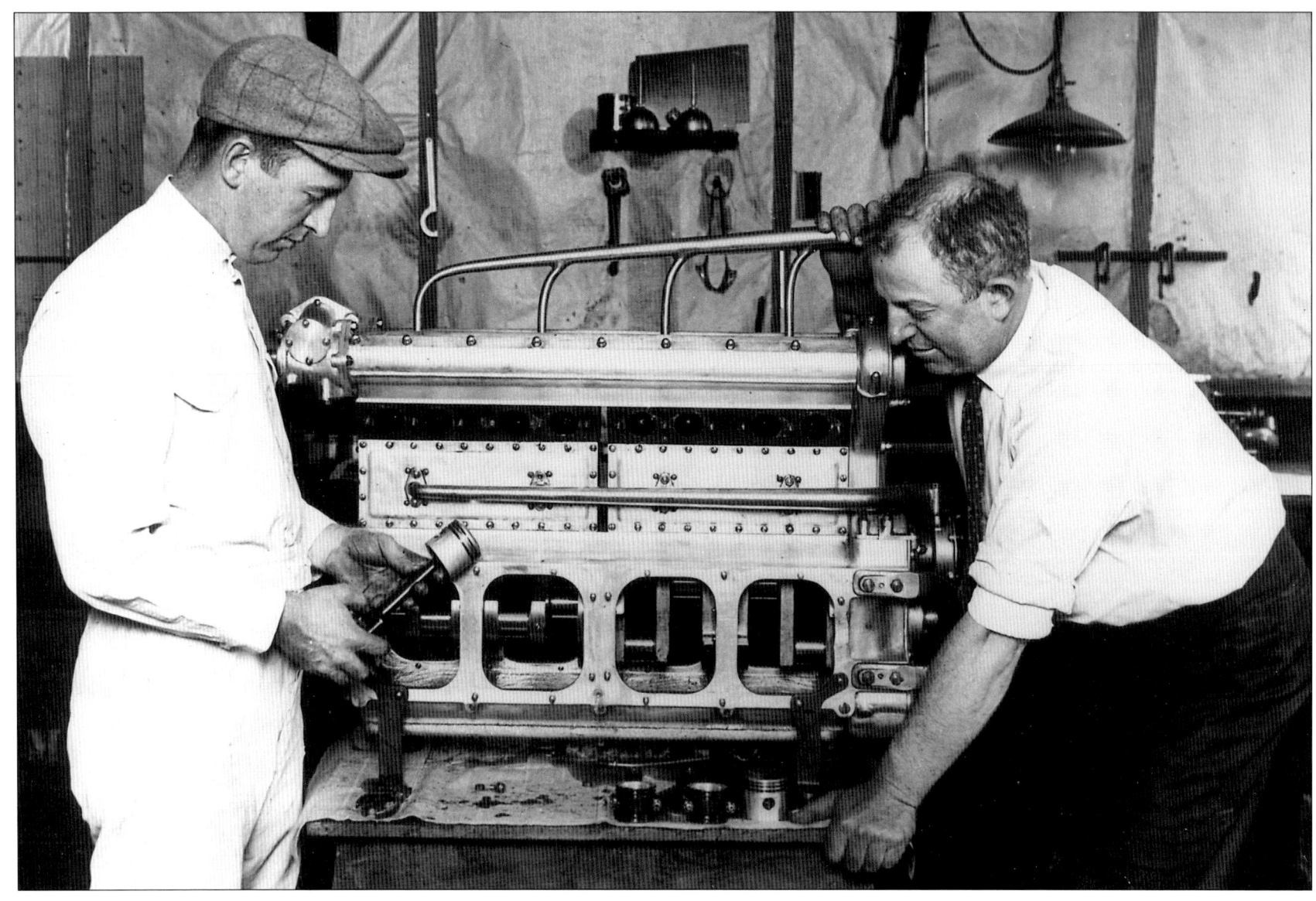

For the 1923 season, Indianapolis and the Contest Board of American Automobile Association reduced the championship class engine displacement to 122 ci, thus outmoding the 183. Designer-draftsman Leo Goossen came into his own with the Miller 122, making refinements that would last in Miller and Offenhauser engines for half a century. He gave the new engine five main bearings and hemispherical, two-valve combustion chambers. In this photograph, Murphy contemplates the tiny 2.334-inch piston while Miller leans on his masterpiece. Murphy used this engine in a Durant chassis to finish third at Indianapolis in 1923.

European racing officials decreed that for 1926, engine displacement would be cut to 91.5 ci and Indianapolis followed suit, thus outmoding the 122 after only three years. Miller had adopted centrifugal supercharging on the 122 and all of the succeeding Miller 91s were built originally as supercharged engines. The engine would become the pinnacle of Harry A. Miller's art. This example was installed in Frank Elliott's 1927 Junior Eight Special.

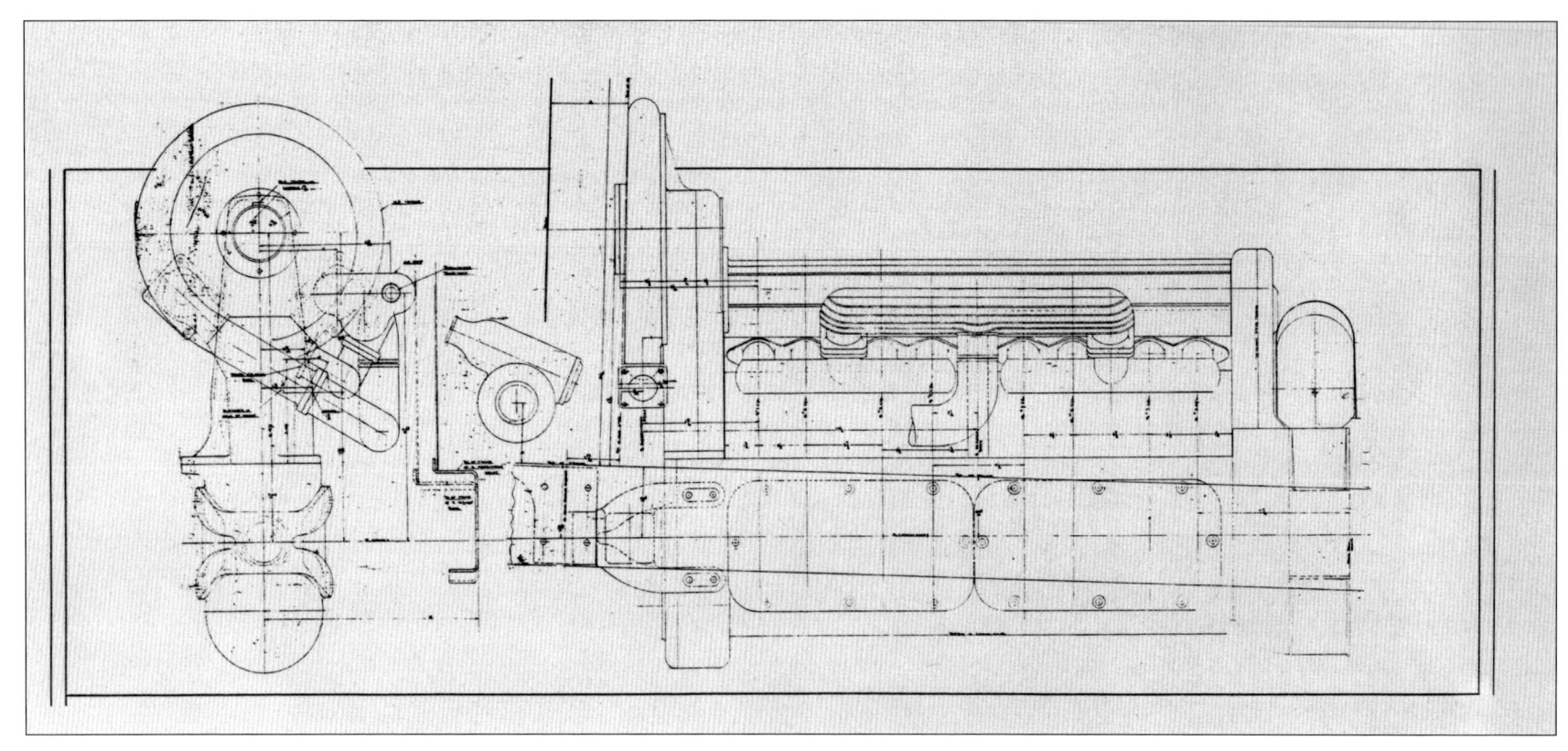

This is the outline drawing of the supercharged Miller 91 of 1928, an engine installed in Peter De Paolo's Flying Cloud Special, driven by Wilbur Shaw.

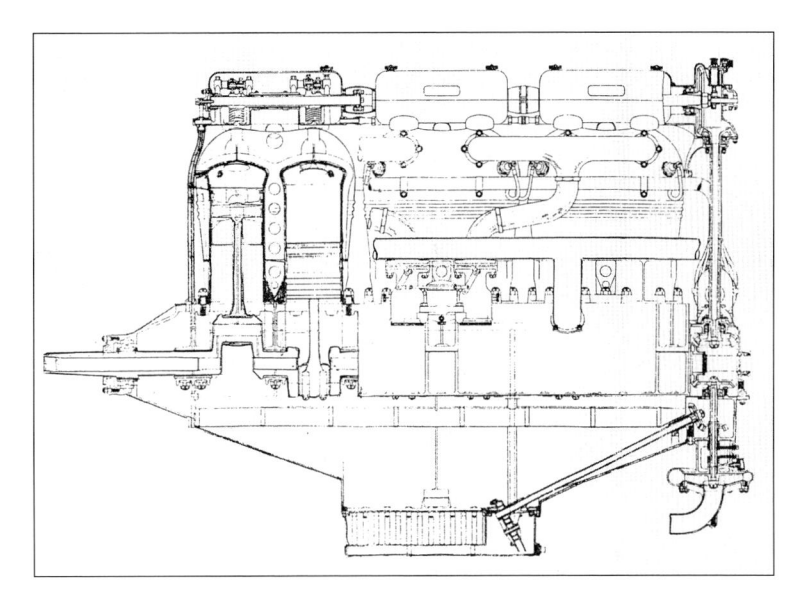

Miller built several aircraft engines, beginning in 1915 with this six-cylinder single-cam 638-ci job for the Christofferson Brothers. The Christoffersons' two planes crashed and nothing further came of the engine.

39

Earle Ovington, the United States' first airmail pilot, backed Miller in his design of a 1,414-ci V-12 aero engine for military use in 1917. A prototype of this large engine was tested at the Army's McCook Field in 1918 just after the United States entered World War I. The engine failed the Army's tests, but was probably doomed anyway by the development of the larger Liberty aircraft engine.

Although built for the small speed boat racing class, the Miller Marine, a 151-ci twin-cam engine, was quickly adapted to auto racing and was very successful on the dirt tracks. This polished example shows its kinship with the 220-ci Offenhauser sprint car engine, although the later Offy was much lighter.

Chapter 4: RACING BOATS

In the early 1920s, American power boat racers were happy to attain speeds of 40 mph. In fact, the inboard hydroplane class record for engines of 151 ci, which produced about 35 horsepower, was just over 36 mph, and was held by Dick Loynes, of San Diego. The shift of auto racing engine design from the pre–World War I monsters of 450 ci to 1923's 122-ci engines led some boat owners to put race car engines in their speed boats. Miller Indianapolis 122 straight-eights easily outclassed the pre-war boat engines and speeds jumped by 50 percent or more.

Loynes was the premier power boat racer on the West Coast and worked with Miller to develop both the un-supercharged 122 and a blown variety that Miller sold for $6,000—12 times the price of the engines the power boat men had been using. The cost led the American Power Boat Association to slap a ban on superchargers and a price cap on boats running in the 151 class. Miller met that challenge with a simplified two-valve-per cylinder unblown marine engine: his 151. This engine spawned a family of four-cylinder racing engines that culminated in the Offenhauser that would dominate American racing for 50 years.

Harry himself had a racing boat, *Angeles I*, in which he installed the first Miller Marine 151. Unsupercharged, it set a mark of 43 mph, then with a blower, turned more than 60 mph, bringing in an immediate demand for Miller engines from the race boat crowd. Within a year, the value of the 151 had also been grasped by auto racing people and the engines were running in cars at the Ascot track in Los Angeles and elsewhere.

In 1926, following the introduction of the supercharged Miller 91 that fit into the International 1-1/2 liter-formula, those engines as well became attractive to the boat racers. Well before Lockhart's 1926 victory in the 500 with a Miller 91, Carl G. Fisher, owner of the Indianapolis Motor Speedway, had Miller convert one of the engines to marine use and raced it on the Thames, in London, for the Duke of York Cup. Even though floating driftwood disabled Fisher's *Little Shadow*, an 18-foot two-step hydroplane built by Purdy in Port Washington, Long Island, it and two other Miller-powered boats showed their speed in England.

In 1926, Miller also upped the displacement of the 122 straight-eight by half, to 310 ci, using blocks of the 151 four. Dick Locke won the 1926 President's Cup in Washington, D.C. with that engine.

The 310 was doubled into a V-16 of 620 ci for Gold Cup competition, with the two banks of eight cylinders joined to the single crankshaft by a "linked" master connecting rod to which the cylinders on each side were joined. It was an arrangement used at the time by Curtis and Packard, but which raised problems of balance and synchronization. Loynes, who would have nothing to do with linked rods, had Leo Goossen design a special gearbox to join two 310s with separate crankshafts. One of the Miller 620s eventually went to Horace Dodge, who won the 1936 Gold Cup in a hull named *Impshi*.

In 1925, Gar Wood advertised Miller outboard motors. Apparently he was anticipating a design that Miller would not get around to building for another three years. In 1928, Miller built an outboard boat engine based on a single-cam engine of 31 ci, but the completed engine lay unused at Miller's Gramercy Street plant until the bankruptcy. It was auctioned with the rest of Miller's fixtures, but disappeared during World War II.

In 1928, Miller built a horizontally opposed-eight-cylinder marine engine of 148 ci. Placed in Jim Talbot's *Miss Riocco III*, the engine appeared to be powerful and the boat fast, but *Miss Riocco III* apparently hit a submerged object and sank. After being raised, the 148 was replaced by a Miller Marine 91. The 148 engine reappeared in California in 1978 while Mark Dees was doing research for his Miller Dynasty book. The late Don Vesco discovered it at a swap meet in 1999.

Talbot, looking to take the Harmsworth Trophy away from Gar Wood, and the water speed record as well, if he could, then commissioned Miller to mount 24 World War I surplus Liberty aircraft engine cylinders on a common crankcase in a "W" configuration. Miller built two of these 3,298-ci monsters, which Talbot installed in a hydroplane he called *Miss Los Angeles II*. The engines performed satisfactorily but *Miss Los Angeles II* capsized in the trophy race at Detroit. The engines were eventually cannibalized and scrapped.

In 1931, Wood, the American speed boat king, was looking for an engine with which to defend the Harmsworth Trophy and the water speed record. His existing Packard marine engine were no match for the latest British Rolls-Royce aircraft engine. At Indianapolis in May he ran across Harry Miller and ordered a pair of supercharged V-12 engines of 1,113 ci that Miller promised him would easily outclass the Rolls model "R" Schneider Cup power plants. Originally to have been a scaled-up 620, it became a multiple of the 220 racing car engine, with four valves per cylinder; its blocks eventually had the same dimensions as the Miller (later Offenhauser) 255.

The engines were to be tested and ready to run in Detroit on Labor Day, an impossible three-month deadline to design, cast, machine, assemble, and test a new engine. However, with Leo Goossen and Miller at work around the clock at a shop in Detroit they came close. A testing accident prevented Wood from using the Miller engines and he was forced to use his old, outclassed Packards, but won the Harmsworth race when Kaye Don's *Miss England II* capsized. The two 1,113s, at last report, belonged to collector Buck Boudeman.

Another engine (which become the best known Miller-powered race engine), a 620, went into John Shibe's *Miss Philadelphia* and then to Zalmon Simmons, of the Simmons Mattress Co. Simmons had Arno Apel, of Ventnor, New Jersey, build a revolutionary three-point hydroplane. The boat rode on just two sponsons and its propeller, which drastically reduced the wetted surface and therefore reduced drag. Simmons' new boat was given the name *My Sin*. (Apel and his father, Adolph, had developed the three-point design for a high-speed boat for the Chinese government. The first race boat they built to the design was *Miss Manteo II*, in 1936.)

Charles Zumbach rebuilt the Miller 620 in New York and discovered that the linked rods gave slightly different stroke length to the left and right banks of the engine. He replaced them with fork and blade rods, which vastly improved the engine's power and smoothness. Thus reworked, *My Sin* won the 1939 and 1941 Gold Cup competitions.

After World War II, Simmons sold the boat to band leader Guy Lombardo who re-named it *Tempo VI*, the sixth in a line of 225 hydroplane race boats and power cruisers he kept at his home in Freeport, Long Island. Lombardo won the Red Bank Sweepstakes in *Tempo VI* in 1946, and the Gold Cup that year on the Detroit River, despite competition from boats powered by war surplus Allison aircraft engines of 1,710 ci. Lombardo was defeated by Danny Foster in the Allison-powered *Miss Peps V* in the Gold Cup race of 1947 on Long Island's Jamaica Bay. The day of the Miller engine in speed boat racing had passed.

Harry himself had a racing boat, Angeles I, which ran the first Miller Marine 151. This is a reproduction of Angeles I, with a 220 Miller engine, built by restorer Chuck Davis.

Dick Loynes' "Twin Eight" 620-ci Miller Marine racing engine for Loynes' Californian. The original Miller 620 was made of two 310 engines doubled into a V-16 for Gold Cup competition, with the two banks of eight cylinders driving a single crankshaft by "linked" master connecting rods. Loynes, who would have nothing to do with troublesome linked rods, had Leo Goossen design a special gearbox to join two 310s with separate cranks.

Dick Loynes in his Miss California (the ex-Quicksilver III), which was powered by a Miller 151 Marine engine racing at Miami in 1928. Loynes ran it with the supercharger and without, depending on the class in which he competed.

Zalmon Simmons, the mattress magnate, and his mechanic, Charles Zumbach, on the Potomac River in Washington, D.C., at the controls of My Sin, the revolutionary three-point Arno Apel hydroplane, powered by Miller's 620-ci V-16. My Sin beat Miss Canada III to win the 1939 Gold Cup in Detroit, and won the 1941 Gold Cup race. After World War II, bandleader Guy Lombardo bought it and had Zumbach rebuild the engine. Lombardo won the 1946 Gold Cup with it under the name Tempo VI.

The 620-ci Miller Marine V-16 Gold Cup engine.

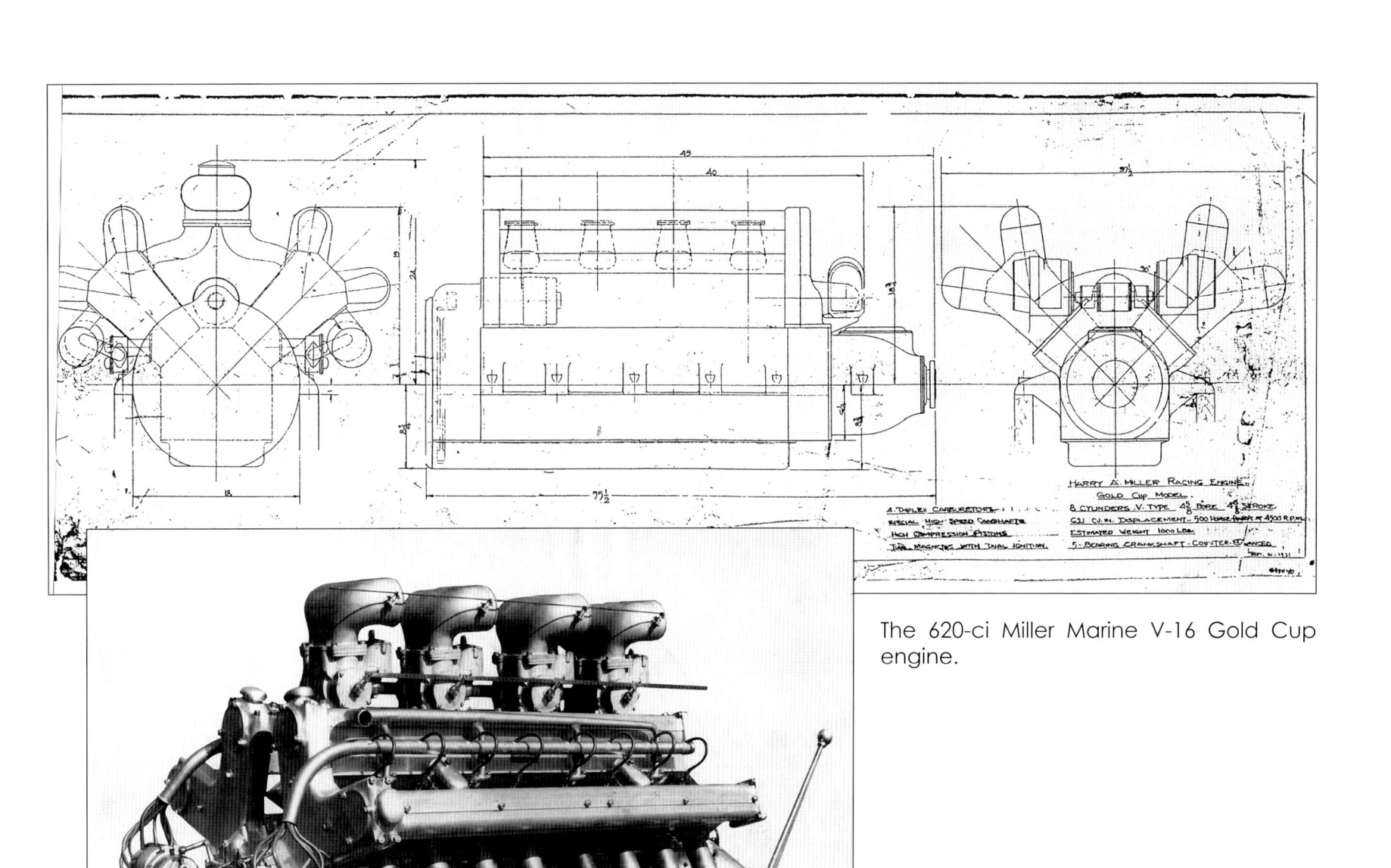

Gar Wood used a supercharged 1,113-ci Miller V-16 engine in Miss America VIII when he attempted to set a water speed record on New York's Harlem River, October 25, 1931. Although Wood and his brother George reached 104 mph, it was not a record. The engine was also run unsupercharged.

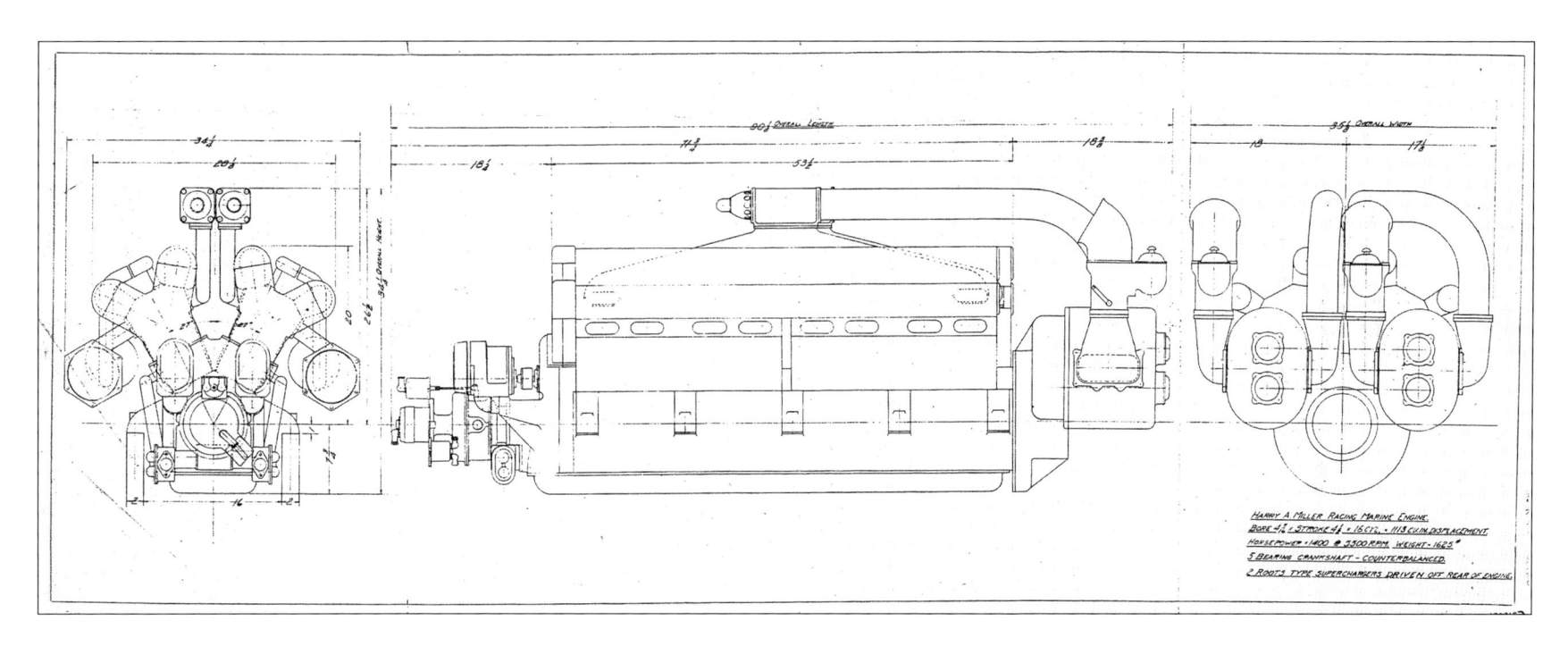

HARRY A MILLER RACING MARINE ENGINE.
BORE 4½ - STROKE 4¼ - 16 CYLS. - 1113 CU.IN. DISPLACEMENT.
HORSEPOWER - 1400 @ 3300 RPM. WEIGHT - 1625#
5 BEARING CRANKSHAFT - COUNTERBALANCED.
2 ROOTS TYPE SUPERCHARGERS DRIVEN OFF REAR OF ENGINE.

Miss Spitfire V, built for George Rand, was powered by a supercharged Miller 91-ci eight. It is shown here off Rye, New York, in 1928. S. C. Chesterton, a Miller draftsman, drew the lines of the 16 and a half-foot hull.

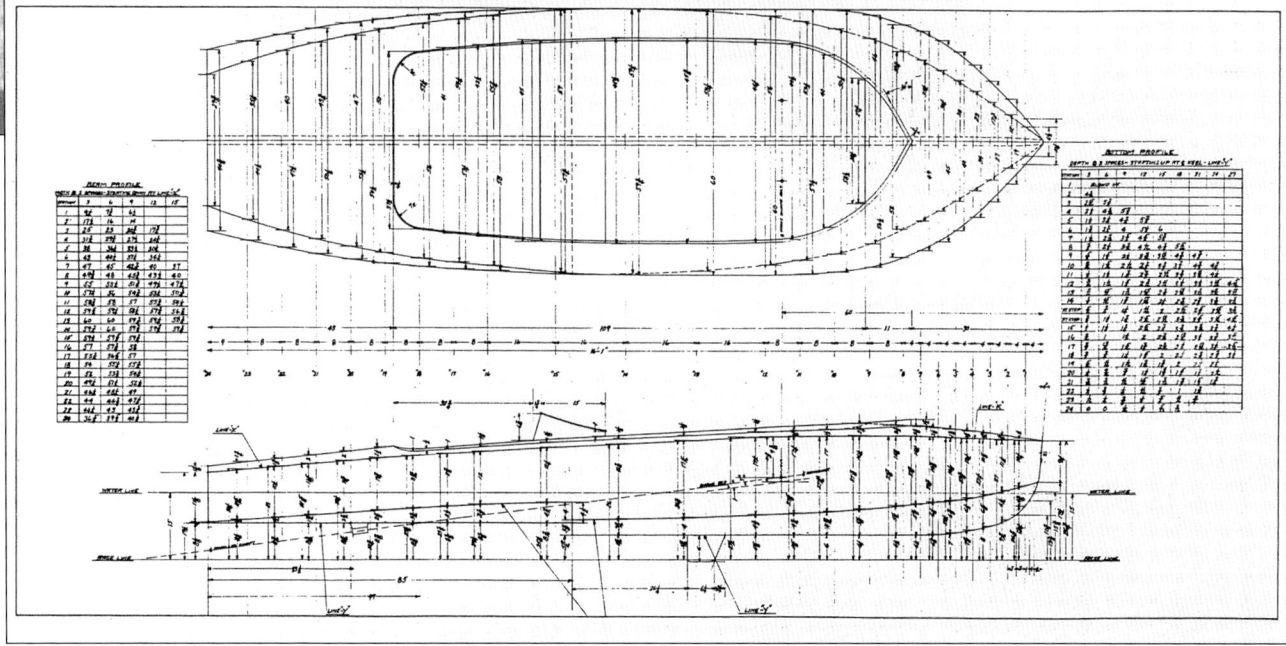

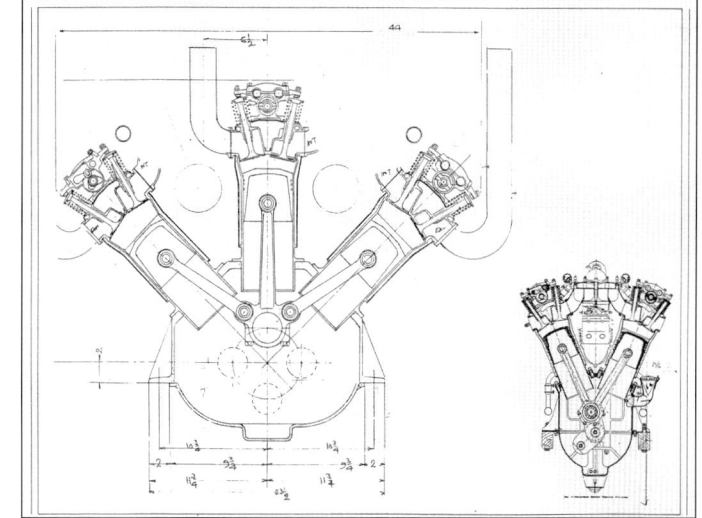

The "W-124" was a monster engine that Miller assembled from 24 World War I surplus Liberty aircraft engine cylinders mounted on a specially cast crankcase. James Talbot of Los Angeles had Miller build two such behemoths and installed them in his Miss Los Angeles II with which he intended to compete for the 1929 Gold Cup. The throttle linkages for the 24 carburetors must have been a week's work to install. Ralph Snoddy drove the Talbot boat at Detroit, aided by two riding mechanics. The boat capsized, which gave Gar Wood the trophy. The engines were later cannibalized to keep Liberty-powered ferry boats to Santa Catalina Island running.

The last marine engine Miller built was this 724-ci V-12, built for Canadian E. A. Wilson in 1936. When Miller was unable to do the necessary development work and to iron out some persistent bugs, Charles Voelker was called in from Detroit to prepare it for the 1937 Gold Cup race. By 1939, Miss Canada III was one of the most powerful boats on the circuit, winning the President's Cup on the Potomac, coming in second in the Gold Cup and winning the APBA championships at Red Bank, New Jersey.

A 122-ci supercharged Miller straight-eight, adapted for race boat use.

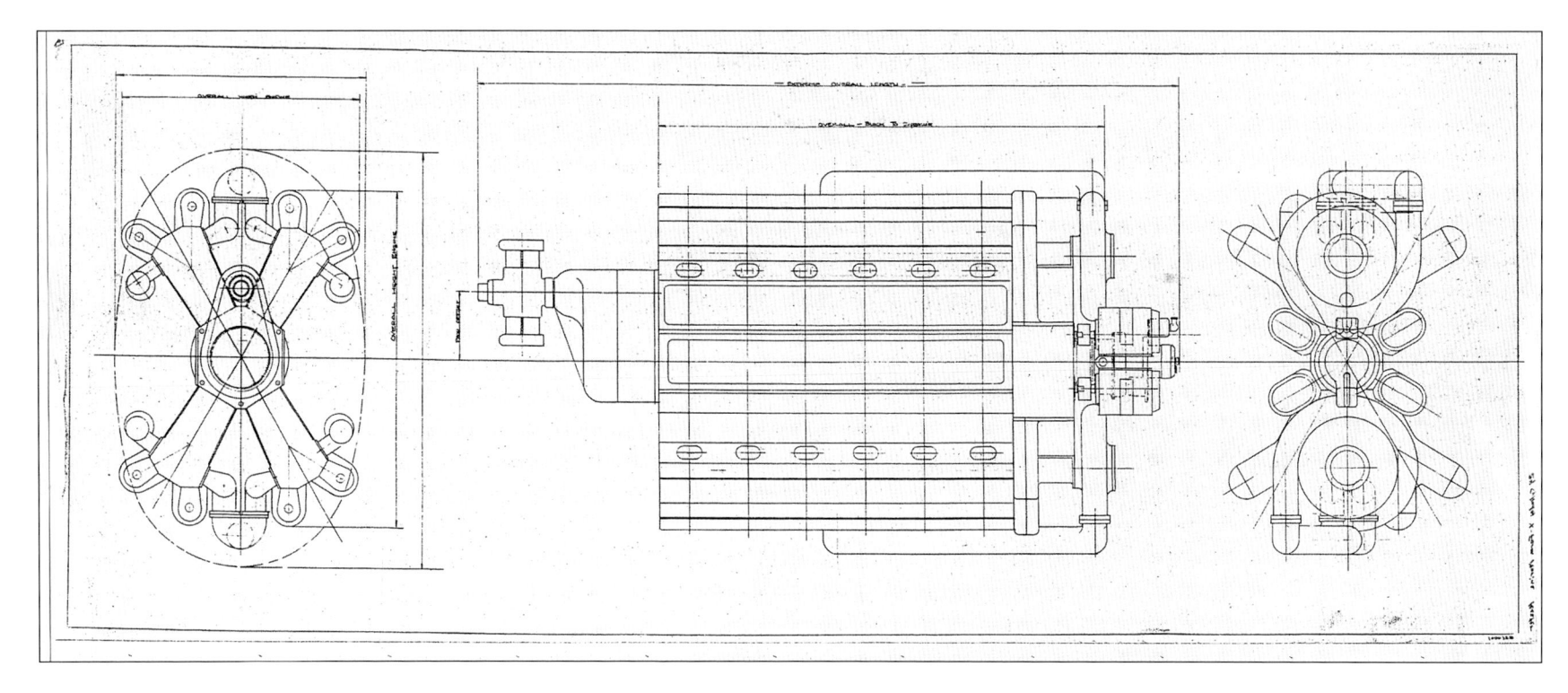

Miller began design studies on this 24-cylinder X-form aviation engine, but work never got to the hardware stage.

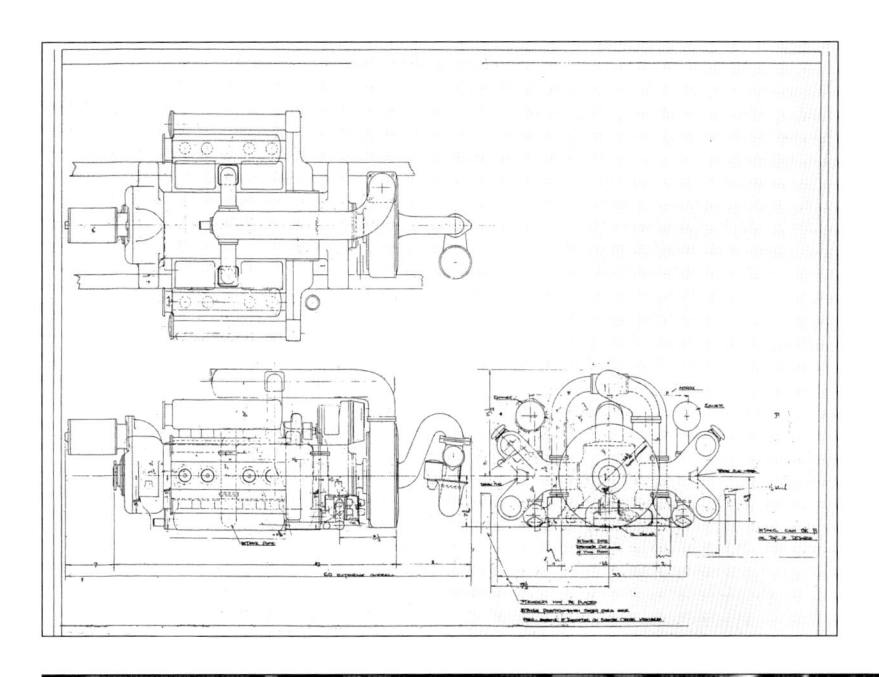

In 1928, at the behest of James Talbot, of Los Angeles, Miller designed this 151-ci opposed-eight four-cam marine engine. Dropped into Talbot's Miss Riocco II, it was significantly more powerful than standard 151 Miller Marine fours. Miss Riocco's hull collapsed at speed and the boat sank. Salvaged, but never run again, the engine sat in southern California until it turned up at a flea market where Miller historian Mark Dees was peddling automotive objects. Bill Manly, of Santa Rosa, has begun to restore the engine to running condition.

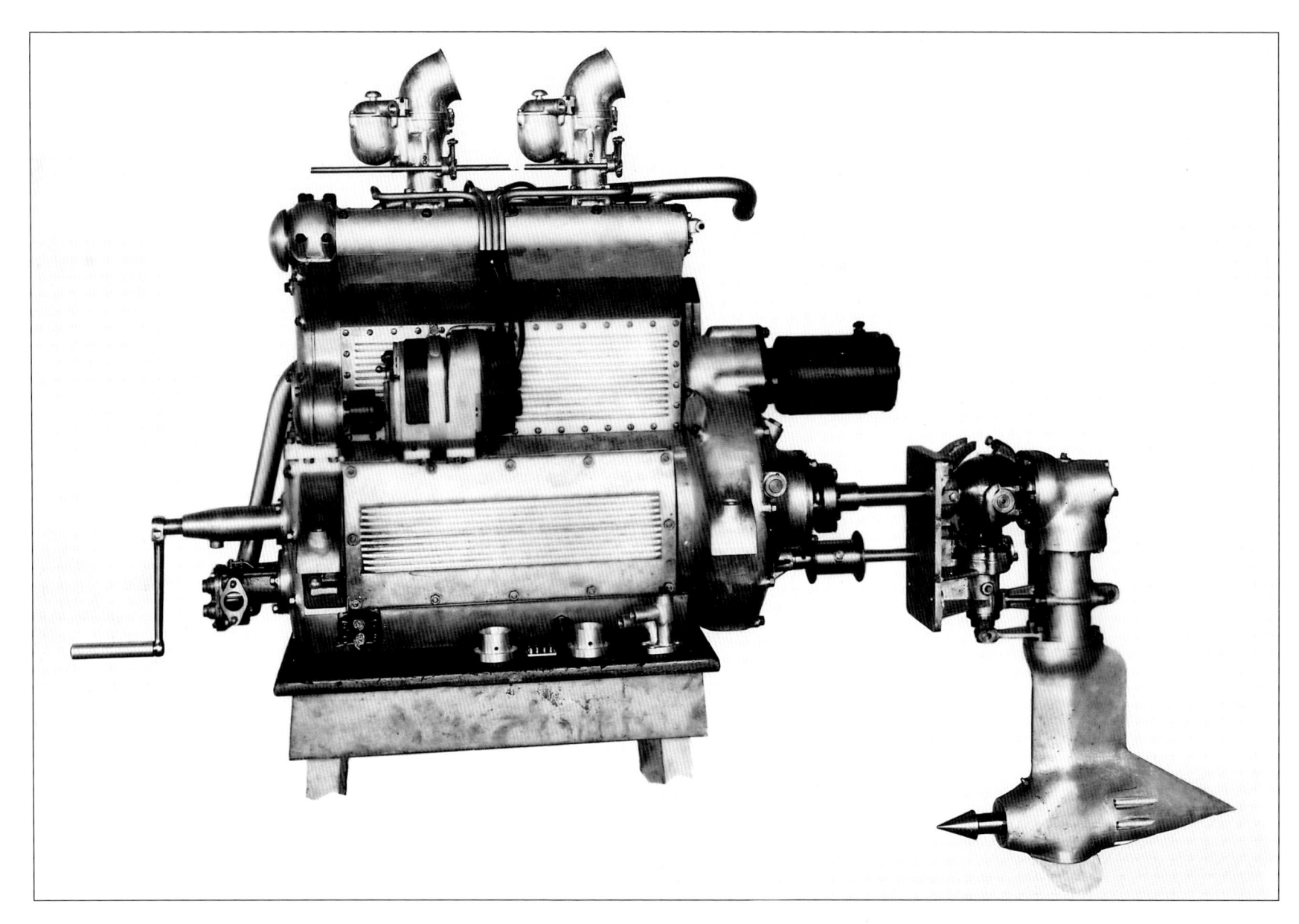

In 1931, Miller designed a marine outdrive that was attached to a 16-valve 151-ci engine. While developed from the bulky 151 Marine engine, this version was much lighter and more tightly designed. It foreshadowed the Miller 200, 220, and 255 engines and through them, the Offenhauser designs that continued to be competitive in American racing into the late 1970s.

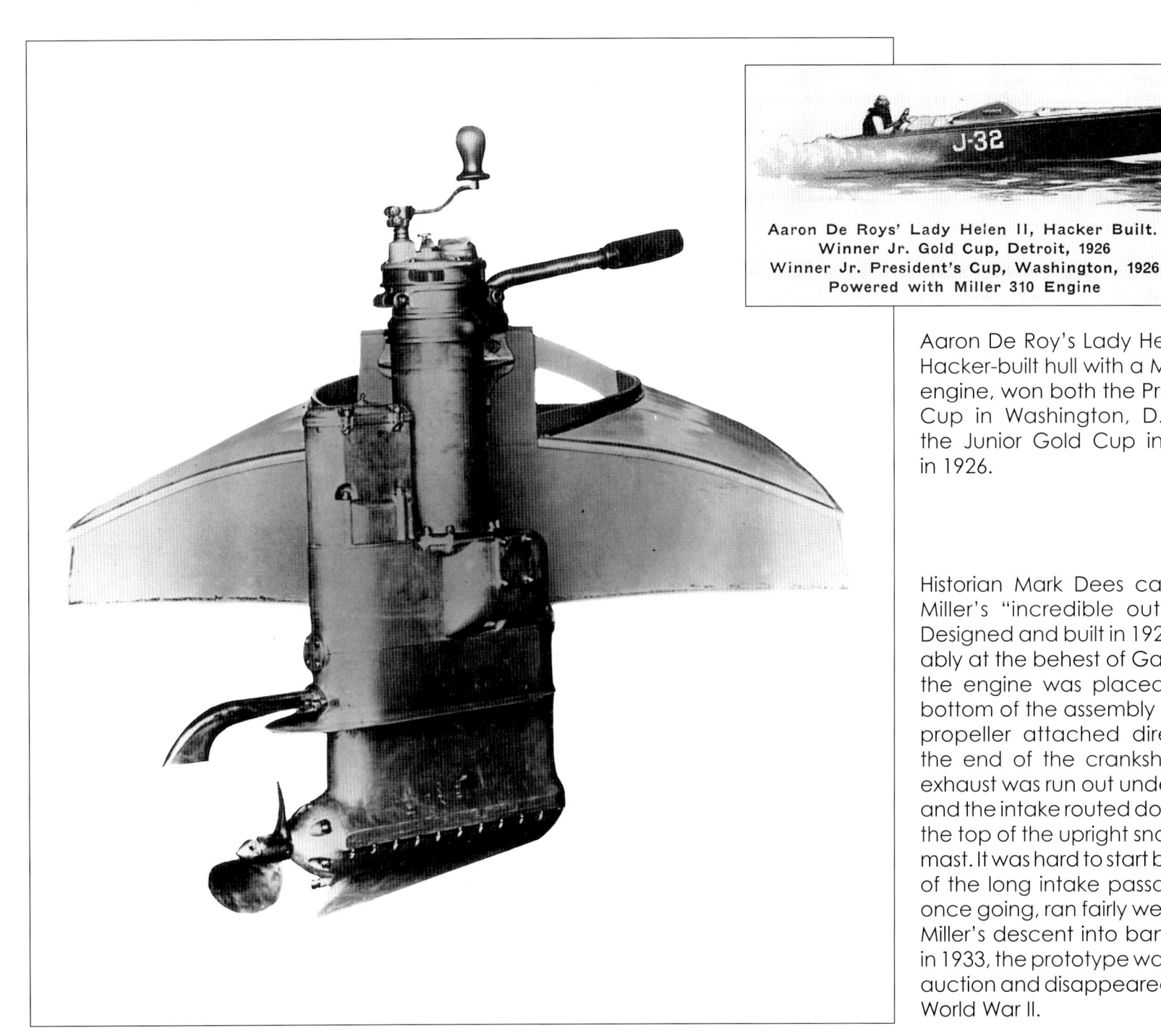

Aaron De Roys' Lady Helen II, Hacker Built.
Winner Jr. Gold Cup, Detroit, 1926
Winner Jr. President's Cup, Washington, 1926
Powered with Miller 310 Engine

Aaron De Roy's Lady Helen II, a Hacker-built hull with a Miller 310 engine, won both the Presidents Cup in Washington, D.C. and the Junior Gold Cup in Detroit in 1926.

Historian Mark Dees called this Miller's "incredible outboard." Designed and built in 1928, probably at the behest of Gar Wood, the engine was placed at the bottom of the assembly and the propeller attached directly to the end of the crankshaft. The exhaust was run out under water and the intake routed down from the top of the upright snorkel-like mast. It was hard to start because of the long intake passage, but once going, ran fairly well. Lost in Miller's descent into bankruptcy in 1933, the prototype was sold at auction and disappeared during World War II.

Chapter 5: PASSENGER CARS

By the early 1920s, Miller was on his way to greatness in racing. However, he would make several attempts as a builder of passenger cars and engines but fall short each time. He conceived far too technically ambitious and complex vehicles.

An engine Miller built for the Leach car was a portent of things to come. The Leach-Biltwell Motor Car Company was organized in Los Angeles in 1918 by a local car dealer. Harry Miller later became an investor in the company. Historian Mark Dees suggests that he may have traded some machine tools from his World War I work on the Bugatti U-16 engine to Leach for stock in the venture. Or possibly, it was just a sponsorship payment for cars run as Leach Specials at Indianapolis in 1921 and 1922.

For 1922, Leach announced a Power Plus model equipped with a 348-ci six-cylinder overhead-cam Miller engine; the car was originally assembled with a six-cylinder Continental engine. Unfortunately, Miller and his designer, Leo Goossen, failed to take into account the second-order torsional vibrations of a six-cylinder engine. All of the Miller engines broke their crankshafts within a few months of service. Replacing them drove the company into bankruptcy. A few mementos of the Leach episode remained in Goossen's effects when he died in 1974.

At the time he was building the Leach engine, Miller conceived a sports car to be modeled along the lines of his racing 122, at the time utterly dominating American racing (six of the first seven places at Indianapolis in 1923). That car never materialized, possibly because of the Leach disaster.

His next excursion into road cars came in 1928, after Miller's front-drive Indianapolis racing car stirred high automotive circles. Errett Lobban Cord bought the rights to Miller's front drive design to use on the L-29 Cord passenger car. While Cord used the Miller name in his advertising, Goossen and Cornelius Van Ranst designed a new, stouter but relatively crude front-drive unit for the passenger car.

Dazzled by Miller's reputation, Phillip Chancelor, a wealthy Californian, asked Miller to build a fast and luxurious custom sports sedan. Racing cars and luxurious road cars are obviously different creatures but even though both Offenhauser and Goossen tried to talk him out of it, Miller agreed. The price was $30,000, probably equal to $500,000 in 2004 dollars. The car would be front-drive, and supercharged.

In order to concentrate the weight as much as possible on the front wheels, Miller and Goossen designed a 308-ci V-8 adapted from the front half of the 620 Miller marine V-16. They used a water-cooled marine centrifugal supercharger that had its own separate radiator. It was engine-turned on all of its surfaces, making it one of the most beautifully finished engines ever manufactured in the United States.

Miller sold his operation to Schofield in early 1929 and the unfinished Chancellor chassis was completed by Offenhauser and Goossen, and then taken to Pasadena to the coachworks of J. Gerard Kirchoff for construction of the aluminum body. Although the car was powerful and fast, the straight-cut engine timing gears were noisy and the car had several teething problems that Goossen, then working for Schofield, had to resolve. Chancellor was disappointed with the car and sold it after about a year. It found its way to William Hollingsworth but through a succession of mishaps, the engine was damaged, leading Hollingsworth to replace the Miller with a 1934 Ford V-8. Hollingsworth eventually sold the car and it appears to have been scrapped some time later.

After Schofield collapsed leaving Miller unpaid for a large portion of what was owed him, Harry returned to Los Angeles and resumed his racing car business. In the summer of 1931, with Indianapolis enthralled by the junk formula and the nation sinking deeper into the Depression, he undertook a new inspiration: four-wheel drive, designed to bring racing traction to its highest possible point. The FWD Company had pioneered all-wheel drive for military trucks during World War I and Miller, the hero of front-wheel drive in racing, would now apply four-wheel drive to the highway.

To power his new creation, Miller had Goossen draw up another 308-ci V-8, but one unlike the Chancellor engine of 1928. The four-wheel drive engine had a split crankcase without the detachable "webs" of his earlier racing engines. (This was a change in Miller practice that he would use in the rest of the engines he designed during his lifetime, including a never-built 1933 aero engine and the Gulf-Miller fours and sixes.)

While developing the racing version, Miller (or Preston Tucker) drummed up two customers for a road version, to use either the V-8 or a V-16 engine. One was William A. M. Burden, a wealthy (of course) New Yorker from Wall Street and the other was Victor Emanuel, a corporation executive from Ohio. The price Burden paid for the V-16 version was $35,000, while the asking price to Emmanuel was $20,000 and the trade-in of a model J Duesenberg.

Miller's creditors, primarily foundries that had not been paid for casting work for more than a year, forced him into bankruptcy in July of 1933, leaving both cars incomplete. Burden hired Goossen, Offenhauser, and Eddie Offutt to finish his car, while Emanuel got only a frame and a few parts. The Burden car was unsatisfactory and in 1938, Burden sold it for $200 to Fred Offenhauser (who was acting for Miller), then working for Gulf at Harmarville, Pennsylvania. Miller planned to rework the V-16 as an Indianapolis engine but never completed the conversion. After Miller's death, Lou Rassey bought it from the estate and put it into a chassis Russ Snowberger had built from Miller front-drive parts. Shorty Cantlon was driving the car, the Auto Shippers Spl, at Indianapolis in 1947 when he crashed to his death.

This is an advertisement for the 1922 Leach passenger car, taken from *Motor Age* for December 1921. The car featured a Miller single-overhead-cam six, but Harry, not understanding the torsional vibrations inherent in a six, did not take them into account; the crankshafts all snapped in a few weeks and the Leach Company went under.

59

In 1928, Phillip Chancellor asked Miller to build him the world's finest, fastest automobile. Harry produced this beautiful front-drive boat-tailed roadster with a supercharged V-8 engine at a price of $30,000 or about $500,000 in 2004 dollars. Pasadena coachbuilder J. Gerard Kirchoff built the body.

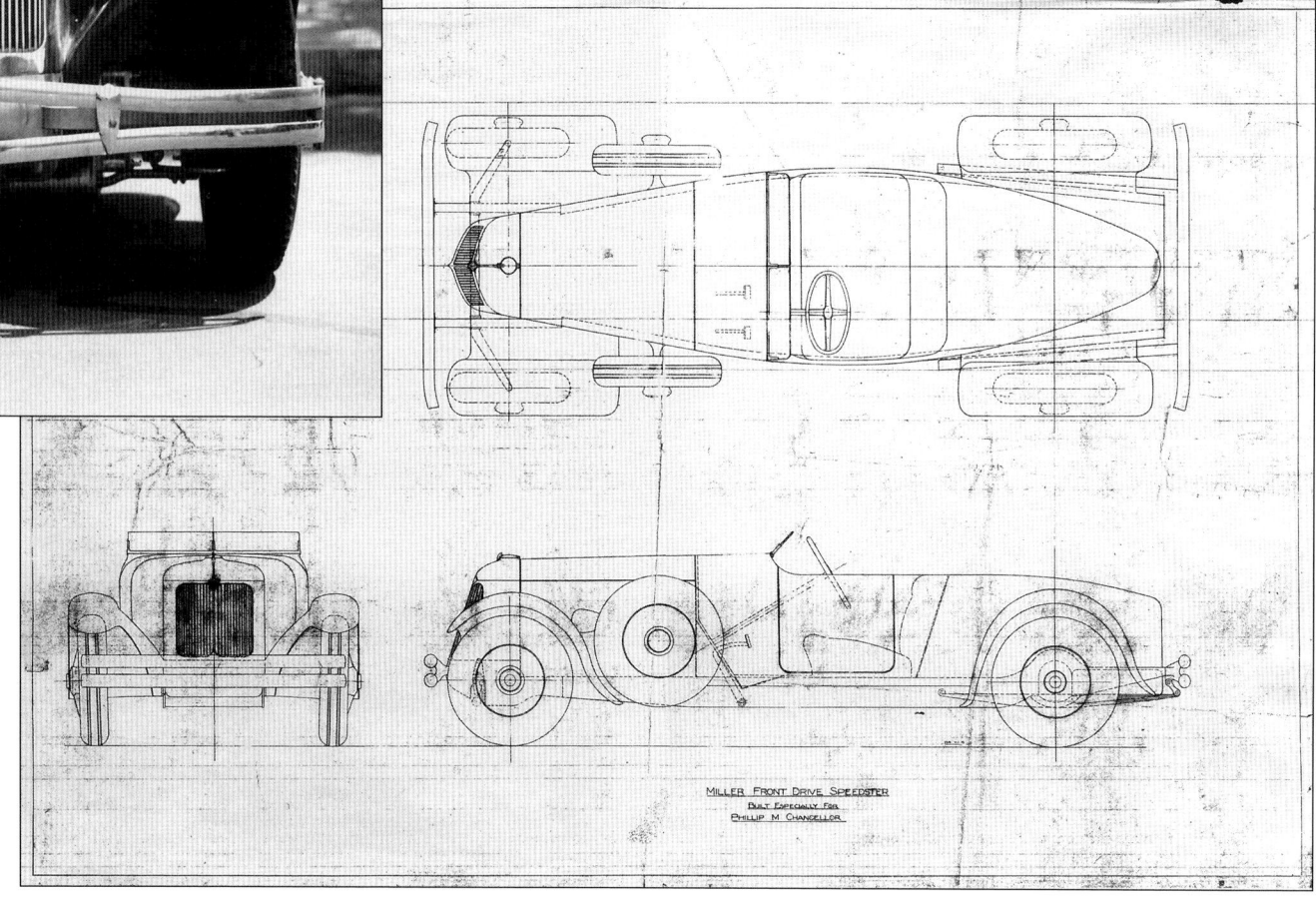

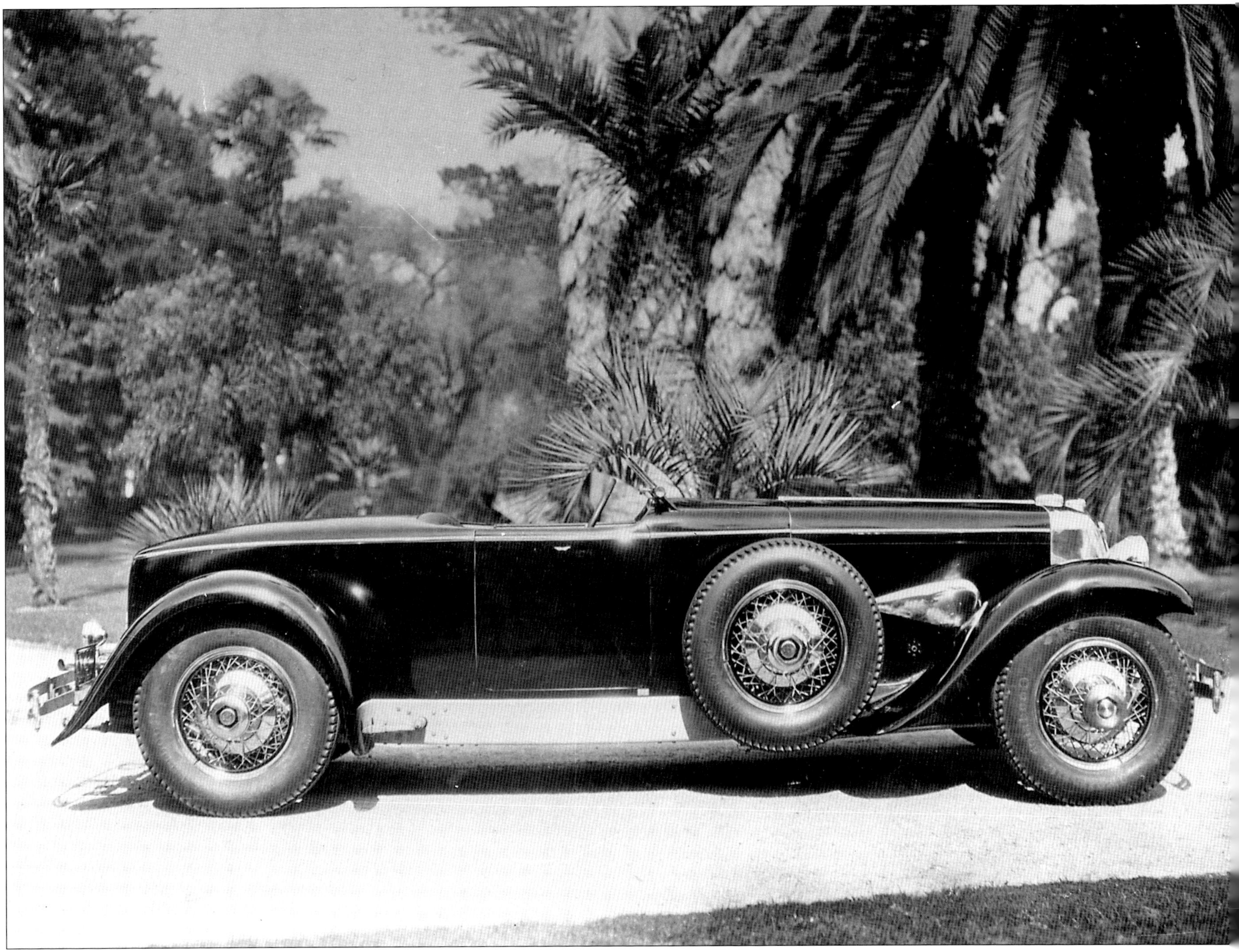

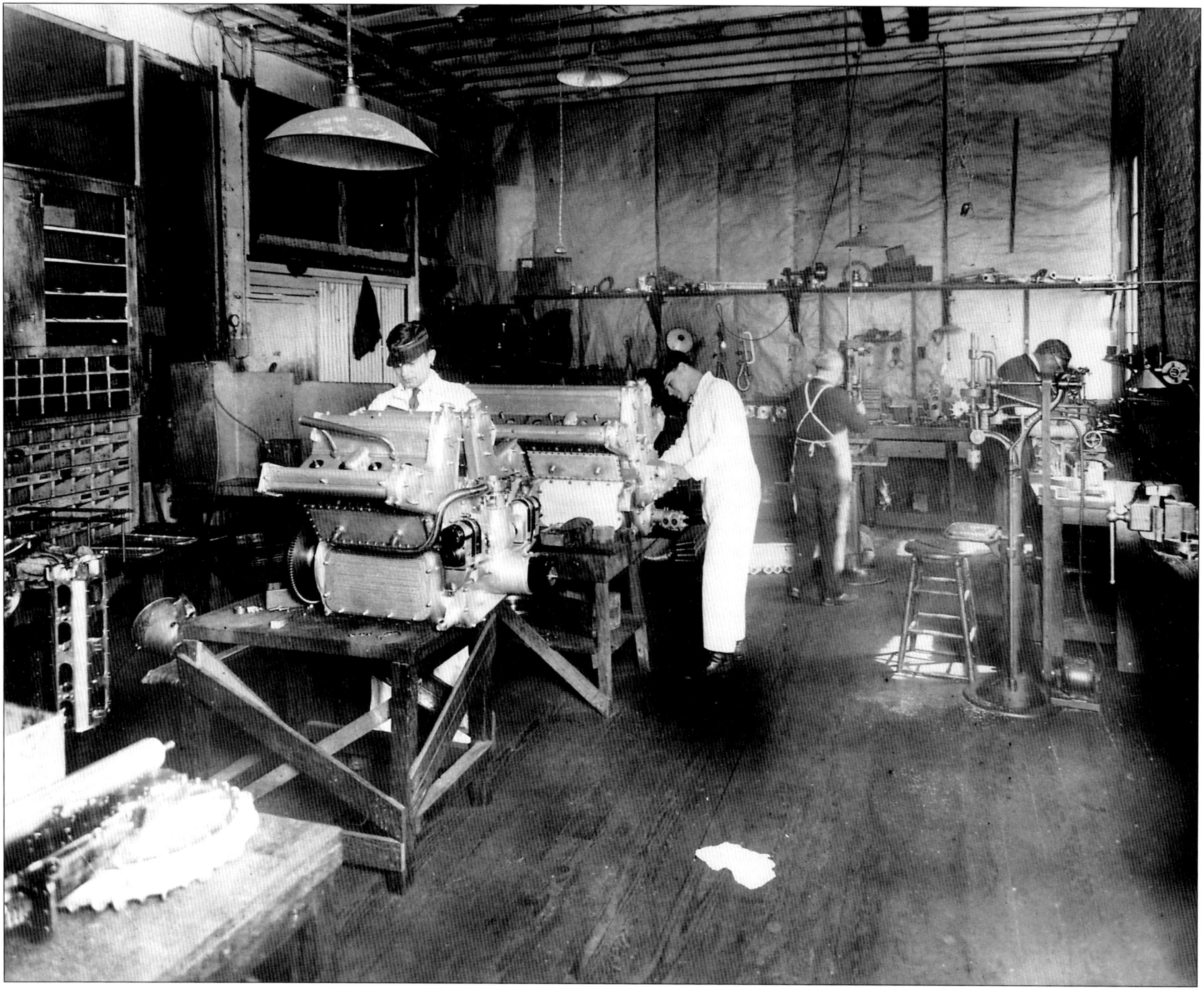

Left: Miller employee Edward Sobraske assembles the Chancellor V-8 while a V-16 speed boat engine goes together behind him. *Above:* The entire surface of the engine was damascened by engine turning.

After Chancellor discovered the car was powerful but neither smooth nor quiet, he sold it and it suffered damage by others several times, eventually coming into the hands of William Hollingsworth, a Los Angeles real estate broker. Hollingsworth removed the beautiful but rough-running Miller engine and replaced it with a Ford V-8 for which Goossen designed a special transmission. The body was updated with a more modern grille and fenders, and eventually scrapped.

In 1930, Miller started work on a replacement engine and transmission for the front-drive L-29 Cord luxury automobile, a wildly impractical scheme, for few could afford such a creation in the depths of the Depression. Cord could not sell their cars with the Lycoming engine, much less a four-cam Miller. This was the prototype, a Cord with a front-drive assembly resembling that of the Miller racing cars.

The engine in the modified Cord, a V-16 of 303 ci, was a beauty to behold.

In 1931, a year after the Wall Street crash, Miller undertook to build the epitome of his art, a four-wheel drive racing car. He managed to sell the idea to the Four-Wheel-Drive Company, as a promotion for their FWD truck line, but Harry carried the idea into a FWD luxury pleasure car. Two customers, William A. M. Burden and Victor Emmanuel, an airline executive, took Miller's bait.

The engine for the Burden car was a 303-ci V-16, similar to that used in the racing car.

In 1938, Burden sold the car to Fred Offenhauser for $400. Fred was acting as a go-between for Miller himself. Although Harry intended to put the engine into a racing car, it was not completed at the time of his death in 1943. Lou Rassey bought the parts from Miller's estate and assembled this engine, put it in an old front-drive chassis, and gave it to Shorty Cantlon to drive at Indianapolis in 1947. Cantlon was killed in the car after a brush with Bill Holland in one of the Blue Crown Specials.

Chapter 6: LAND SPEED RECORD CARS

Running for world unlimited speed records was a lucrative and a not too risky a business early in the twentieth century. Barney Oldfield became a household name for his speed runs. As tires and chassis improved, high speeds were attained with few crashes. World War I, with its accelerated development of aircraft engines, lifted the sport to higher and more dangerous levels. The availability of large, cheap surplus aircraft engines led to drivers hooking up the largest engine they could find, or two or three engines, in an unsophisticated frame, and seeing how fast they could go. Englishman Parry Thomas mounted a 1,649-ci American Liberty V-12 in Count Louis Zoborowski's Brooklands car, Babs and took it to more than 170 mph on Wales' Pendine Sands in 1926. A year later, trying to regain his title after Sir Malcolm Campbell had upped the record to 174 mph, Thomas was killed when Babs crashed. H. O. D. Seagrave raised the record to 203 mph at Daytona in early 1927 using a pair of 435-horsepower Sunbeam engines.

American Frank Lockhart, a superb self-taught engineer and racing driver won the Indianapolis 500 as a rookie in 1926 in a rear-drive Miller 91. After much success on the board speedways, he set his sights on the world speed record. He drove his Miller 91 171 miles an hour at Muroc Dry Lake in California for a class record. Calculating that with better streamlining and double the power he could relatively easily beat Seagraves' record. Lockhart persuaded the Stutz Company to sponsor an attempt.

Lockhart, Myron Stevens, Floyd Dreyer, John and Zenas Weisel and other experienced race car fabricators built the Stutz Blackhawk car at the Stutz plant in Indianapolis. Apparently superbly streamlined, it was powered by two supercharged Miller 91 engines hooked together on a common crankcase in a U-16 configuration. Several features of the car, after the fact, look dubious. It used, for advertising purposes, a Stutz worm-gear rear axle drive, said to be subject to lock-up on the overrun when the driver lifted his foot off the gas at speed. Although the separate wheel fairings had slightly less frontal area, they had much more surface area than an envelope body would have had. A 40-gallon fuel tank that weighed 300 pounds when full was installed in the tail of the car—even though the car would run fewer than ten miles at a time.

On April 25, 1928, after a crash from which he emerged only slightly injured, Lockhart made a run, northbound against the wind, of 193.1 mph and a return run of 203.5 mph. While attempting another run northbound, a tire failed and he crashed to his death. (Aerodynamicists later suggested that the car's tail heavy design put the center of effort ahead of the center of gravity, making it unstable in a crosswind.)

Riley Brett bought the wreckage of the Lockhart car and, in 1938, when the AAA again allowed superchargers at Indianapolis, he and Alden Sampson built a modern chassis to carry it. The engine seemed ideal for the 1939 race where the rules permitted a supercharged 183. Brett's driver, Bob Swanson, qualified third fastest, but dropped out due to driveline problems. A succession of oiling problems and the intervention of World War II prevented the combination from achieving what appeared to be its potential.

In the spring of 1932, Barney Oldfield, another promoter and self-promoter, got together with Harry and proposed a land speed record car designed to break Campbell's 253.970-mph mark. It was to have been powered by a supercharged 3,000 horsepower V-24 engine made up of Miller's marine engine components. It would have three Roots-type blowers in a 180-inch wheelbase chassis covered with an envelope type body. (Wheels, rotating forward, create vast amounts of drag, greatly exceeding the frontal area and skin drag coefficient of a body that covers them.)

Deep in the Depression, the Oldfield car remained a pipedream that never went past the talking and drawing stage, even though Miller thought for a time about developing a record car capable of 200-mph speeds for hours at a time on endurance runs.

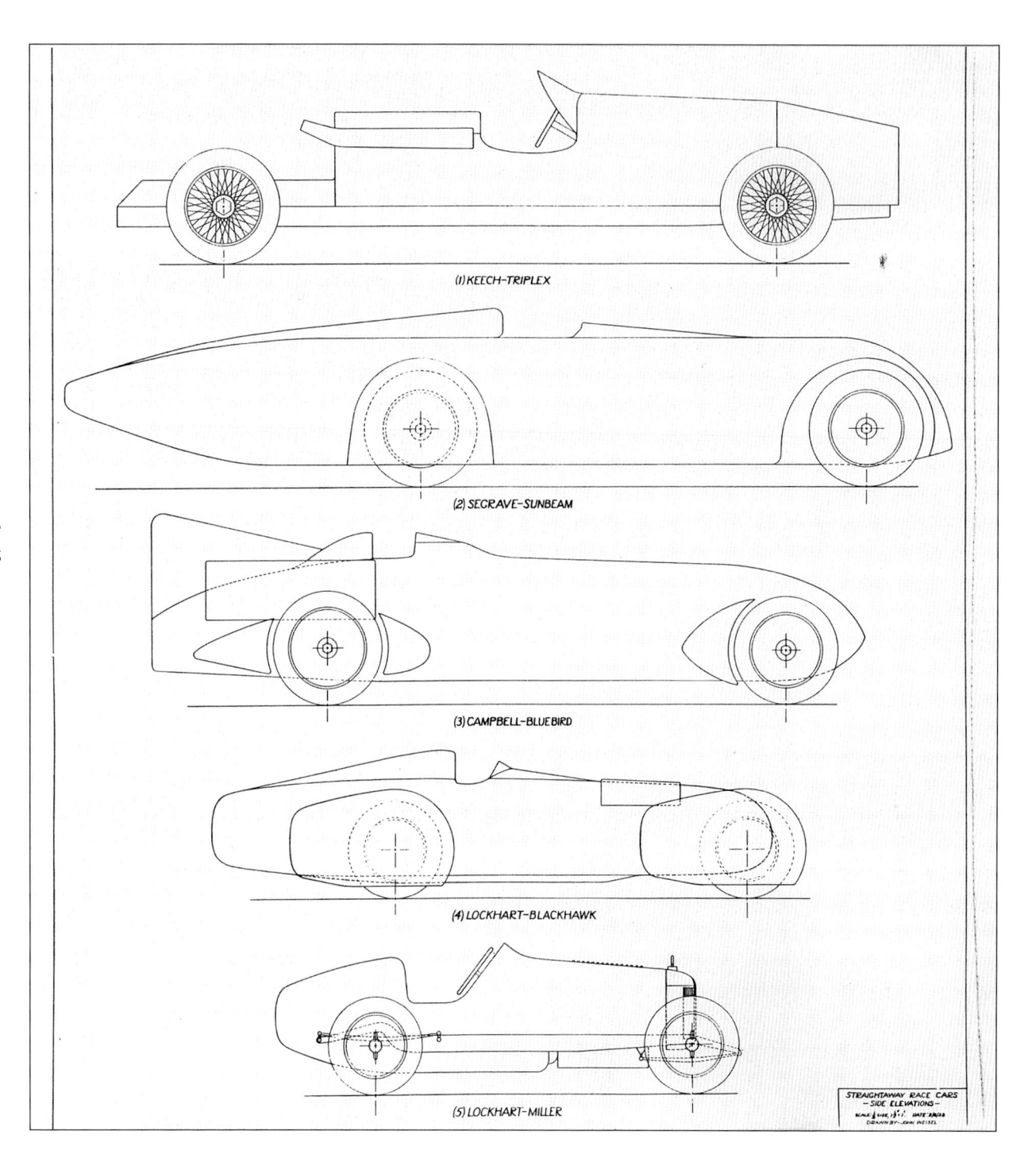

A comparison of some of the cars that ran for speed records during the 1920s.

(1) KEECH-TRIPLEX
(2) SEGRAVE-SUNBEAM
(3) CAMPBELL-BLUEBIRD
(4) LOCKHART-BLACKHAWK
(5) LOCKHART-MILLER

Young Frank Lockhart burst upon the American racing scene like a meteor, coming off the California dirt tracks to win the Indianapolis 500 as a rookie in 1926 in this Miller rear drive 91.

In April 1927, Lockhart ran for a class record at Muroc Dry Lake, California, going one way at 171 mph in his Miller and averaging 164.28. A month later in the same car he turned a 147.72-mph lap at the Atlantic City board track, at that time the fastest oval track lap in history.

STUTZ BLACK HAWK SPECIAL 1928 LSR ATTEMPT.

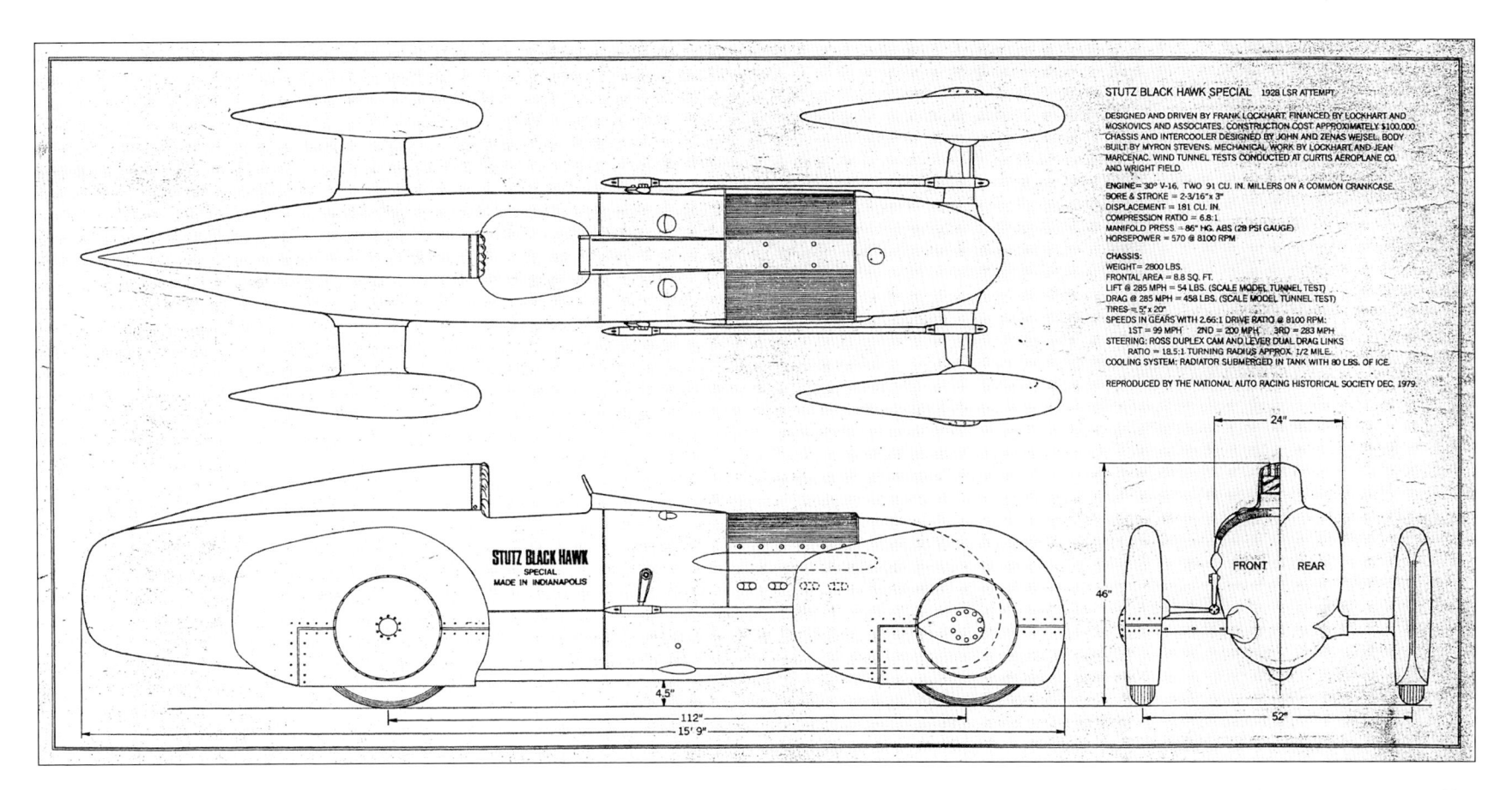

DESIGNED AND DRIVEN BY FRANK LOCKHART. FINANCED BY LOCKHART AND MOSKOVICS AND ASSOCIATES. CONSTRUCTION COST APPROXIMATELY $100,000. CHASSIS AND INTERCOOLER DESIGNED BY JOHN AND ZENAS WEISEL. BODY BUILT BY MYRON STEVENS. MECHANICAL WORK BY LOCKHART AND JEAN MARCENAC. WIND TUNNEL TESTS CONDUCTED AT CURTIS AEROPLANE CO. AND WRIGHT FIELD.

ENGINE = 30° V-16. TWO 91 CU. IN. MILLERS ON A COMMON CRANKCASE.
BORE & STROKE = 2-3/16" x 3"
DISPLACEMENT = 181 CU. IN.
COMPRESSION RATIO = 6.8:1
MANIFOLD PRESS. = 86" HG. ABS (28 PSI GAUGE)
HORSEPOWER = 570 @ 8100 RPM

CHASSIS:
WEIGHT = 2800 LBS.
FRONTAL AREA = 8.8 SQ. FT.
LIFT @ 285 MPH = 54 LBS. (SCALE MODEL TUNNEL TEST)
DRAG @ 285 MPH = 458 LBS. (SCALE MODEL TUNNEL TEST)
TIRES = 5" x 20"
SPEEDS IN GEARS WITH 2.66:1 DRIVE RATIO @ 8100 RPM:
   1ST = 99 MPH   2ND = 200 MPH   3RD = 283 MPH
STEERING: ROSS DUPLEX CAM AND LEVER DUAL DRAG LINKS
   RATIO = 18.5:1 TURNING RADIUS APPROX. 1/2 MILE.
COOLING SYSTEM: RADIATOR SUBMERGED IN TANK WITH 80 LBS. OF ICE.

REPRODUCED BY THE NATIONAL AUTO RACING HISTORICAL SOCIETY DEC. 1979.

When Lockhart decided to go for the world speed record, this is the car he designed and had built. One of the most beautiful racing cars ever made, it was far more esthetically pleasing than the contemporary behemoths that ran for unlimited records in the post World War I era. Unfortunately, it had at least five flaws, two of which may have contributed to Lockhart's death. It used the Stutz worm gear rear axle, which tended to lock up when the driver lifted off the gas. In addition, aerodynamically, its center of effort was ahead of its center of gravity so that it was unstable in a side wind. It also had a 40-gallon fuel tank, which was unnecessary for a record car and the short exhaust stacks probably cost power, as they had no scavenging effect. The carburetor intakes, which were flush with the car's skin, were also a failure as the airflow characteristics starved the engine until they were changed.

After a minor crash in his Black Hawk Special, Lockhart had gotten up to 225 mph at Ormond Beach, Florida, when a gust of wind or a blown tire sent the car into a sickening slide and a series of flips, destroying the car and killing Lockhart.

This diagram, by an American Automobile Association Contest Board investigating committee, shows how Lockhart's Blackhawk skidded, dug into the sand, and flipped.

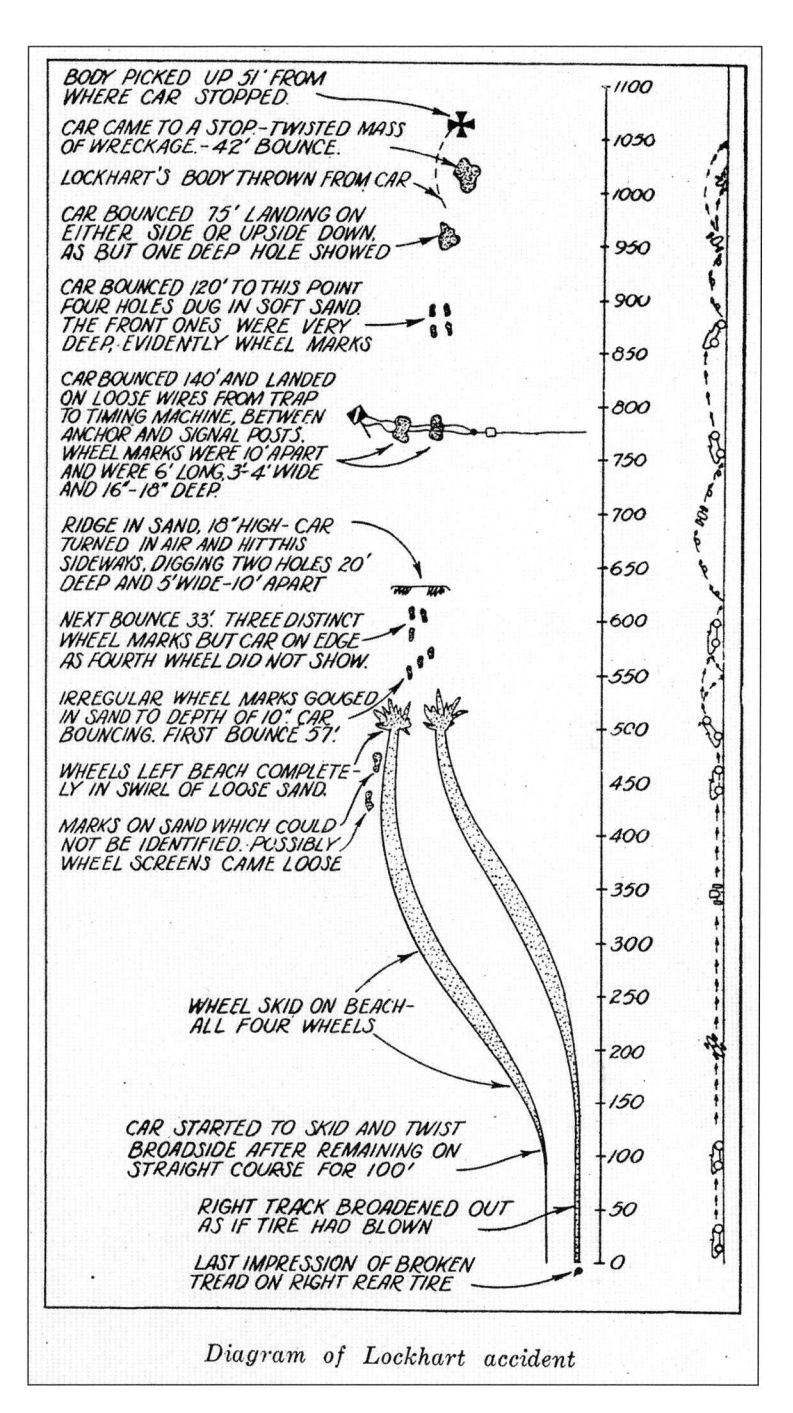

Diagram of Lockhart accident

Riley Brett bought the remains of the Lockhart car and salvaged the engine. In 1929, he installed it in an Indianapolis car that qualified, but lost its oil during the race. In 1938, Brett rebuilt it as the Sampson Special. The disassembled engine is seen here, displayed in all of its glorious complexity by Alden Sampson.

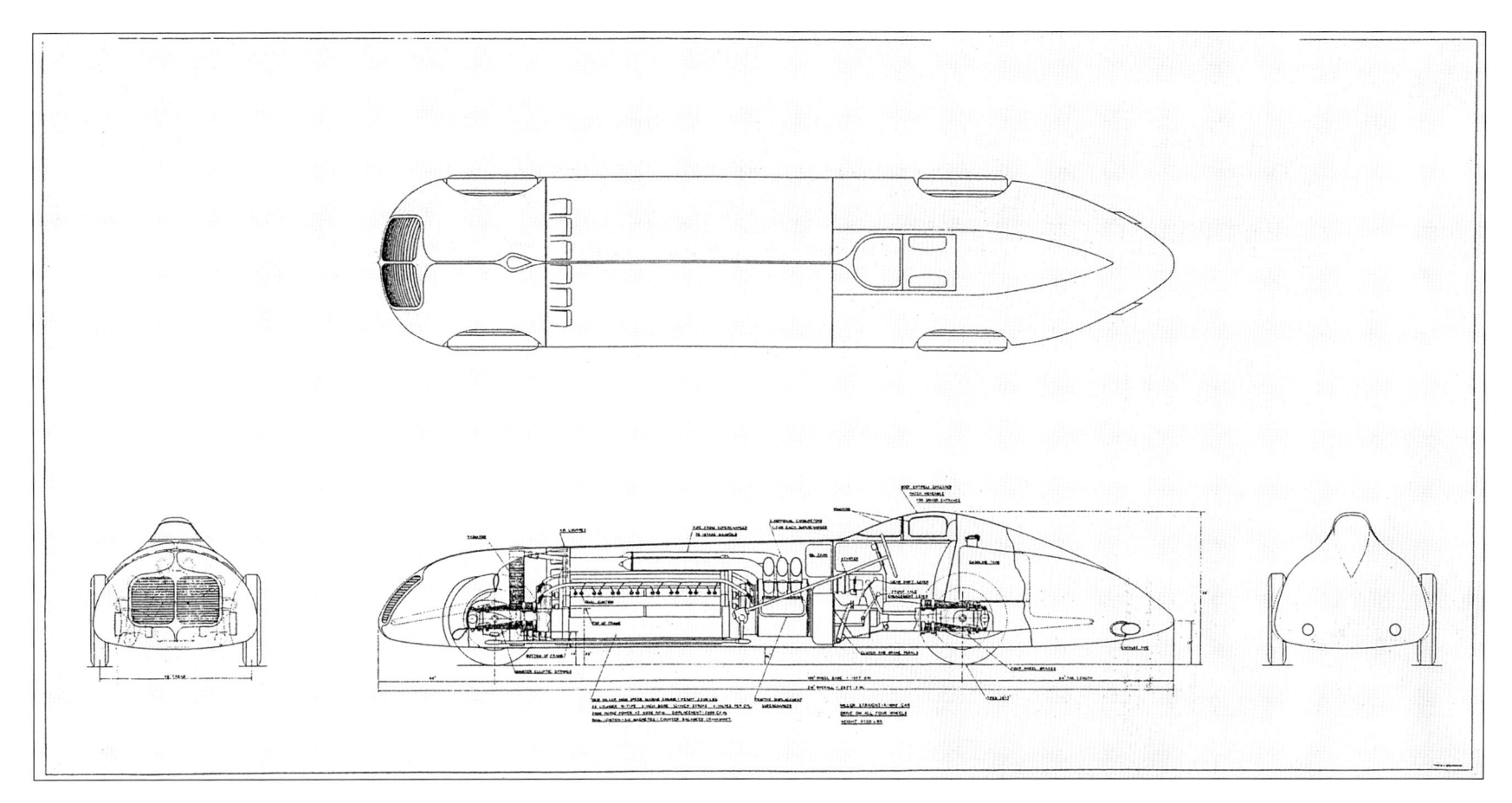

Early in 1933, Harry Miller and Barney Oldfield collaborated on a plan to challenge for the land speed record with this streamlined monster. Money to build it never appeared and within months, Miller was declared bankrupt and left California permanently.

Chapter 7: LATER RACING ENGINES 1930-'33

Despite deep financial difficulties in his final years in Los Angeles, Harry Miller, with the help of Leo Goossen and Fred Offenhauser, designed no fewer than nine new racing power plants, five of them aftermarket heads for the Ford Model A block and four of them complete engines. Three of those engines would become Harry's most lasting legacies, the foundations for the engines that would continue the Miller dynasty as the Offenhauser.

Harry sold the Miller Engine & Foundry Co. to Schofield Inc. in 1929, eight months before the stock market crash of that year. It was also shortly before the Contest Board of the AAA changed the championship rules to what would become known as the "junk formula," of unsupercharged 366-ci engines and two-man chassis. When it took over Miller's business Schofield built parts for Miller's existing Indianapolis cars and engines and produced two heads that Miller and Goossen had designed for the new Ford: a rocker-arm and a twin-cam. Both were immediately taken up by the hot rodders who ran on the streets of Los Angeles and the dry lakes courses. Schofield went bankrupt in 1931 and those designs were bought by Harlan Fengler and Craney Gartz. These two created the Cragar Corporation and renamed the heads "Cragars." The heads stayed in production under the auspices of Roy Richter's Bell Auto Parts Co. into the 1950s. When Schofield went bankrupt their payments to Miller ceased and Harry went back into business for himself, at first under the appellation "Rellimah" and then simply as Harry A. Miller Inc.

Schofield also took the Miller Marine 151-ci four-cylinder eight-valve racing engine and bored its cylinder out about .030 inch to get a displacement of 183 inches. This was a pretty good dirt track engine, and it did well enough at Indianapolis to bring Shorty Cantlon home second in 1930, and generated new interest in four cylinders after a decade of eight-cylinder dominance. However, the 151/183 had been designed for speed boats and was not only a bit heavy at 440 pounds, but had a relatively heavy crank, rods, and pistons.

Goossen was still at Schofield in 1930 so Miller had Walter Steele draw up a flathead head for the Model A and a single chain-driven overhead-cam head for the four-cylinder Ford. After Goossen returned to Miller's fold he drew a new four-cylinder 16-valve engine designed for the cars running at the Legion Ascot track in Los Angeles. It was similar to the 151 Marine, only in having four cylinders and overhead cams, it was 115 pounds lighter, had four valves per cylinder, had lighter rotating and reciprocating parts, and displaced 220 ci. That engine would be used in sprint cars for a generation and became the basis for a slightly larger engine, the 255 Miller, and for the Offenhauser engines that dominated American racing for nearly another 50 years.

As what was planned to be Miller's planned bread and butter championship engine, Harry, Leo, and Fred designed another unblown 16-valve engine of 230 ci, known as the Big Eight. Four chassis and seven or eight engines were built before Harry A. Miller Inc. went under in 1933. The most successful of the Big Eights was Lou Meyer's engine that won at Indianapolis in 1936.

The last race engine to come from Goossen's pencil and the Miller shop was a 182-ci engine with two valves per cylinder for Harry Hartz. Goossen did almost all of the design work, as Miller was heavily involved with his FWD project at the time. In the incredible period of less than six weeks, Hartz, Goossen, Offenhauser, Jerry Houck, and Jean Marcenac drew, cast, machined, assembled, and tested the 182, then took it to Indianapolis in 1932 where Fred Frame qualified it 27th and won the 500.

Much of the 182 architecture was, by necessity, similar to the Miller 220. The engine, as was Miller practice, used two four-cylinder cast iron blocks mounted inline on single aluminum crankcase. Each block had a displacement of 91 ci. Two years later when Earl Gilmore went to Fred Offenhauser to have an engine built for the new midget racing cars, Goossen simply took the front half of the Hartz engine, bored it out a bit to 97 ci, and presto, had an engine that became as nearly immortal as things mechanical are ever likely to be. More than 500 were built and they utterly dominated midget racing until, 30 years later, much larger engines were allowed to compete in midget circles.

When Miller sold his business to the Schofield Company in 1929 that company planned to continue producing Miller racing engines but were confounded by the AAA Contest Board's change in the racing formula to large, 366-ci unsupercharged engines. To feed the market for much cheaper speed equipment, Schofield had Leo Goossen design a twin-cam head for the Ford Model A block. When the Ford crankshafts proved fragile, a heavier, five-main bearing block was cast, to become known as the "D. O. Cragar," seen here as Russ Garnant does some maintenance on his engine.

When Miller returned to Los Angeles after Schofield's bankruptcy, he had Goossen design a cast iron single-cam head for the popular Ford Model A and B engines.

In 1932, Miller had Goossen and Offenhauser design the last successful Indianapolis engine to bear his name, the Hartz 182 straight-eight. Whipped out in an incredible six weeks, the engine was an improved version of the old Miller 122, consisting of two 91-ci four-cylinder blocks on a common crankcase, as had long been Miller's practice in all of his straight-eights. In May, it powered driver Fred Frame to an Indianapolis victory.

HOWDY WILCOX and CHASSIS 1933

To meet the size requirements of the new unsupercharged "junk" formula, Miller designed a 230-ci engine, known as the Big Eight, in 1931. Seen here is Howdy Wilcox with the ex-Marks Miller equipped with a Big Eight.

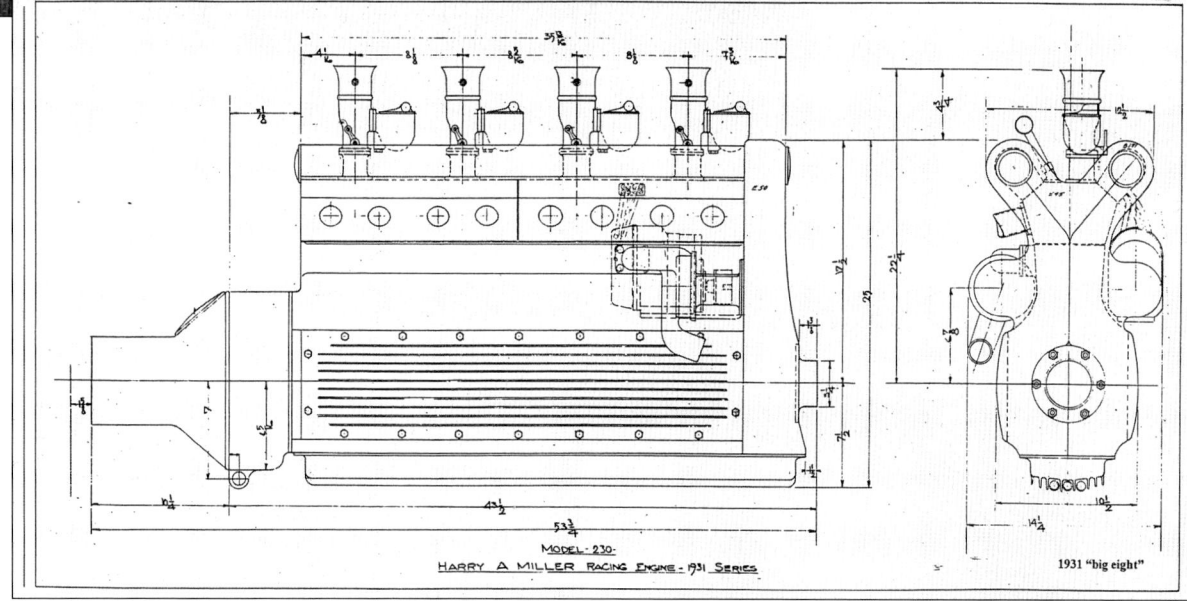

MODEL-230-
HARRY A MILLER RACING ENGINE - 1931 SERIES

1931 "big eight"

Hartz is seen here with the huge Schebler Indianapolis Victory Trophy. The engine, although it was a one-of-a-kind, would live on for years as the basis for the Offenhauser midget engine that Fred Offenhauser brought out in 1934.

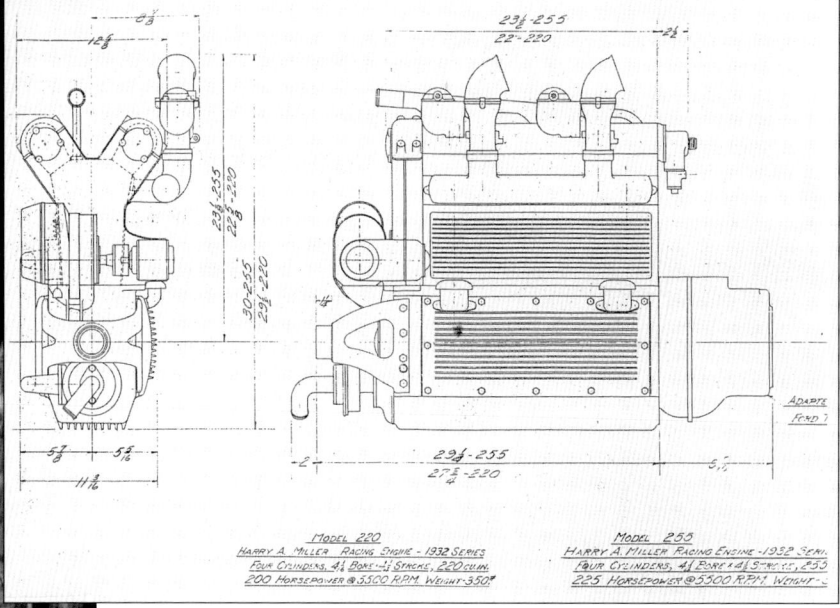

To meet the challenge of the hopped up Fords on the dirt tracks, Miller brought out a lighter version of the 151 Miller Marine, first at 200 ci and later at 220. A slightly enlarged version that displaced 255 ci soon followed. The 220-ci engine was an immediate success at Ascot and in boat racing. This is an early version with the "sidewinder" magneto that was subject to gasoline dripping from the carburetors—not a healthy situation.

One of the 220 versions went into Ted Horn's Haskell Miller. It had downdraft intake passages in the head, which required this odd intake manifold.

85

This is the four-wheel drive car Miller built for FWD truck company sponsorship. The car's best outing at Indianapolis was a fourth by Mauri Rose in 1936. Later, after World War II, Bill Milliken took the FWD and developed it as a hill climb car, and set a new record at the Giant's Despair Hill near Wilkes Barre, Pennsylvania, against far more modern competition and also won at Burke Mountain and Mount Equinox.

The FWD's four-cam V-8 Miller engine, made, uncharacteristically, with a split crankcase.

Chapter 8: EXPERIMENTS THAT FAILED

In 1932 and 1933, while he was struggling to build two racing cars for the Four-Wheel-Drive Company, of Clintonville, Wisconsin, Miller was also attempting to recoup his fortunes by peddling a unique, luxurious road car to William Burden, as told in chapter five. His imagination running away with him, Miller conceived and had Leo Goossen draw up a succession of other innovative road car designs. Leo, who usually considered drafting and design work a pleasant challenge, was only too eager to put pencil to paper for whatever Harry wanted to contemplate. Among sketches that still exist among Miller's drawings and Leo's papers are a dozen exotic custom cars that, after Miller's bankruptcy, no one was willing to buy. They include a January 1933 teardrop-shaped design that pre-dated Buckminster Fuller's Dymaxion and William Stout's Scarab.

After the Miller-Ford disaster at Indianapolis in May 1935, Harry met Thomas Hibbard, a well-known designer of custom automobiles and a former partner in the Lebaron custom coach building company. Hibbard had just left Harley Earl's General Motors Design Studio. He and Miller joined forces with the idea of building a light sports car along the lines of the Miller-Ford chassis. Even though Miller and Hibbard worked on a number of sketches, their ideas never jelled.

Next followed Miller's involvement with Roy S. Evans who had bought the defunct American Austin Car Company, of Butler, Pennsylvania. Evans believed that in those Depression times a small car like the British Austin Seven might be salable in the United States, even though the American Austin company had recently failed. He hired Miller-Hibbard to design engines and bodies for a new version of the Austin to be called the American Bantam.

Miller, Hibbard, and Everett Stevenson moved their establishment from New York to Butler and while working there, Miller proceeded to build a second version of his 723-ci V-12 marine engine, as described in chapter four. Stevenson drew plans for a 71-ci single-cam four-cylinder engine to power the Bantam and Hibbard-designed bodies, but both were far too ambitious for Evan's facilities. Miller was allowed to rent space at Butler for a time, while Hibbard left to become styling head of Ford.

In 1937, before Miller left Butler for the Gulf facilities at Harmarville he designed and built a diminutive four-wheel drive utility vehicle that he drove around the Bantam plant. It is intriguing to contemplate what might have transpired for Miller had he stayed at Bantam. Between 1938 and 1940, Evans produced nearly 7,000 of the little Bantams, slightly restyled from the British Austin.

With World War II approaching, Evans competed for a U.S. Army design for a small, four-wheel drive military vehicle. The result was the Jeep, which in many ways resembled Harry Miller's 1937 plaything. Bantam won a contract to produce the original jeep, but the company was judged by the War Production Board to be incapable of building enough of the vehicles to meet the military's needs, so the bulk of the contract went to Willys-Overland, and eventually to Ford.

This wash drawing was the proposed Miller-Hibbard sports roadster, to be built on a Ford V-8 chassis, but it never got past the drawing board.

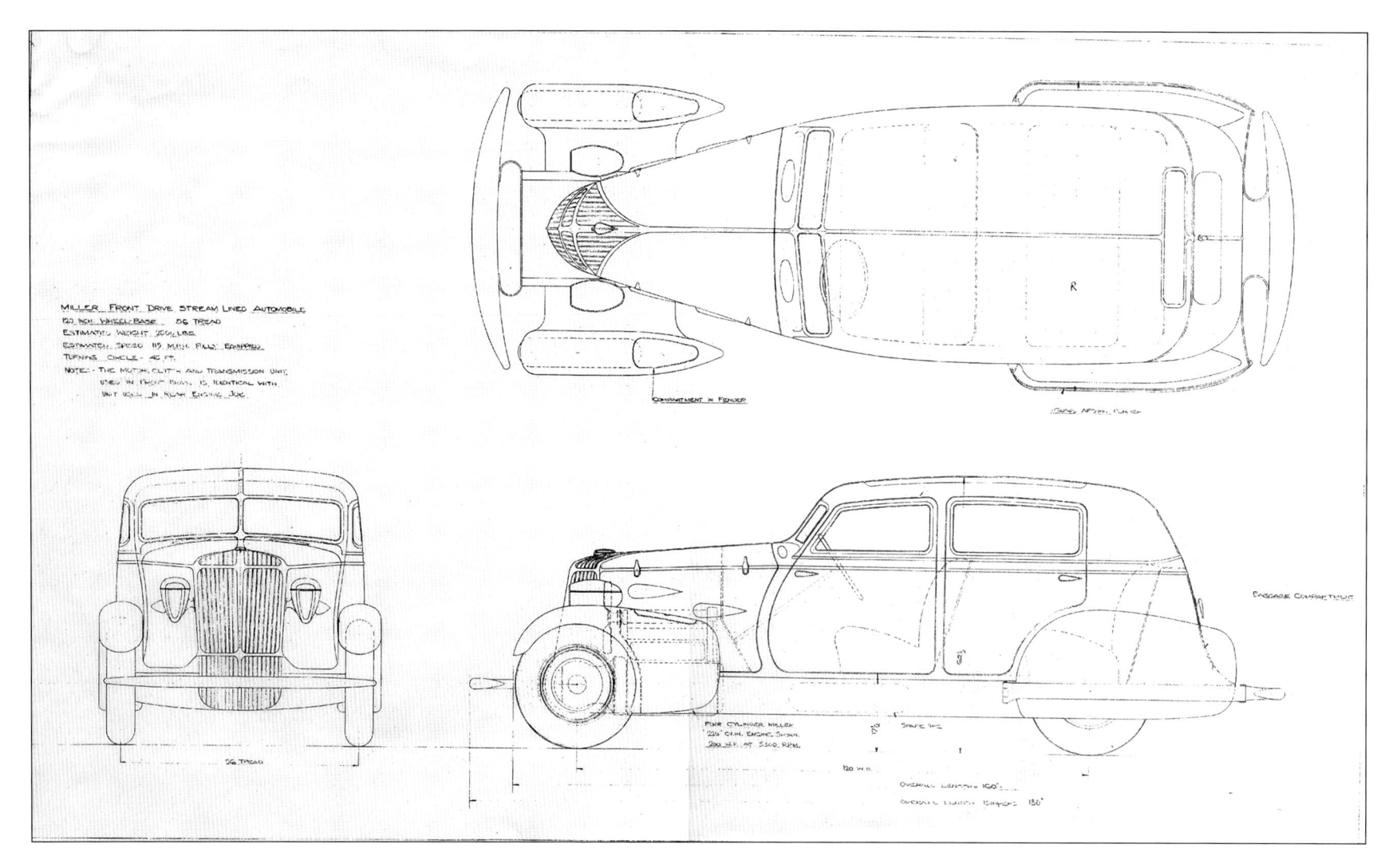

Leo Goossen's drawing for a proposed "streamlined" V-8 front-drive passenger car. The nearest the concept came to reality was the V-16 car built for William A. M. Burden.

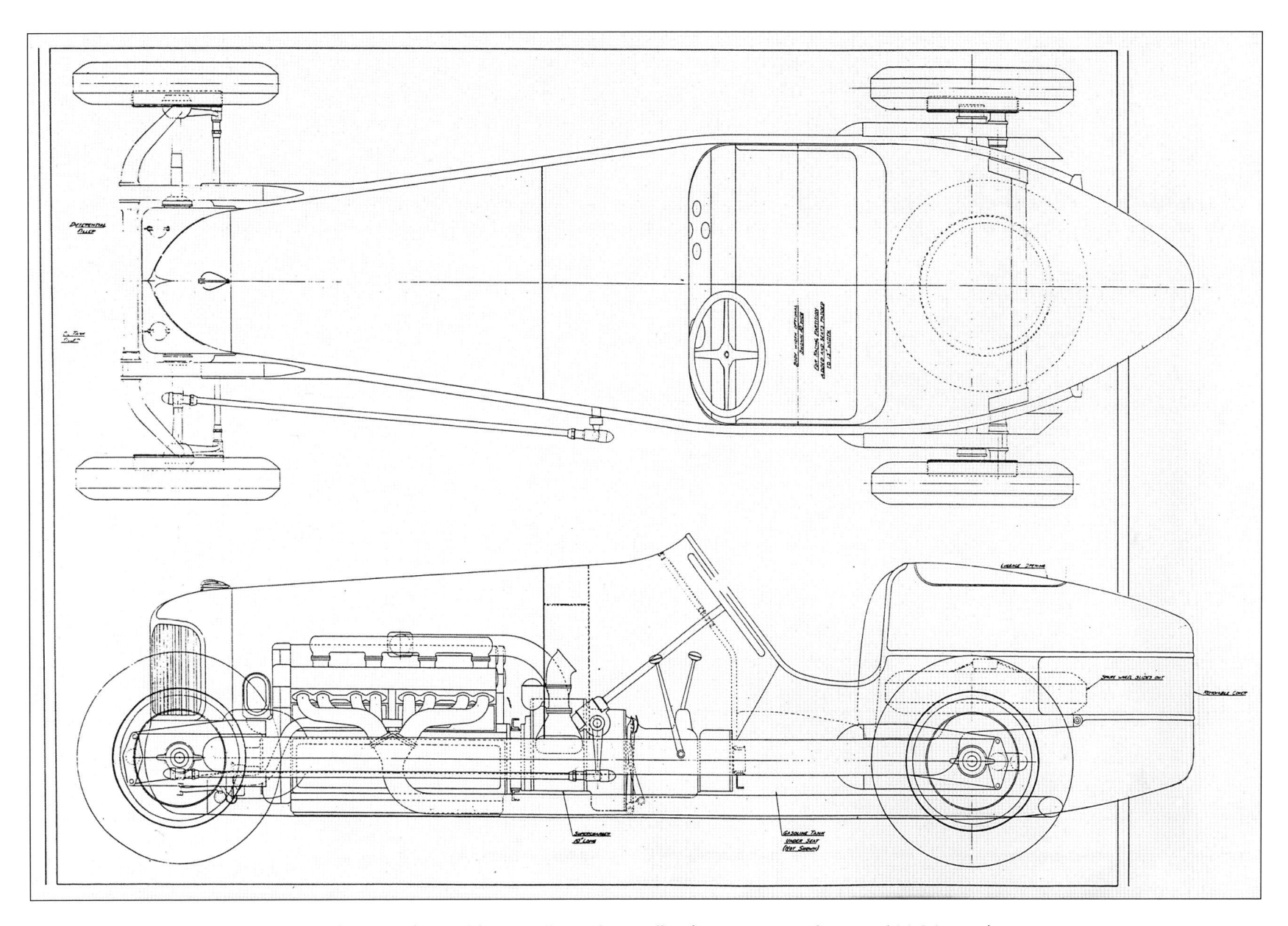

This looks like a race car but Miller designed it as a "sports car," using a supercharged V-16 engine.

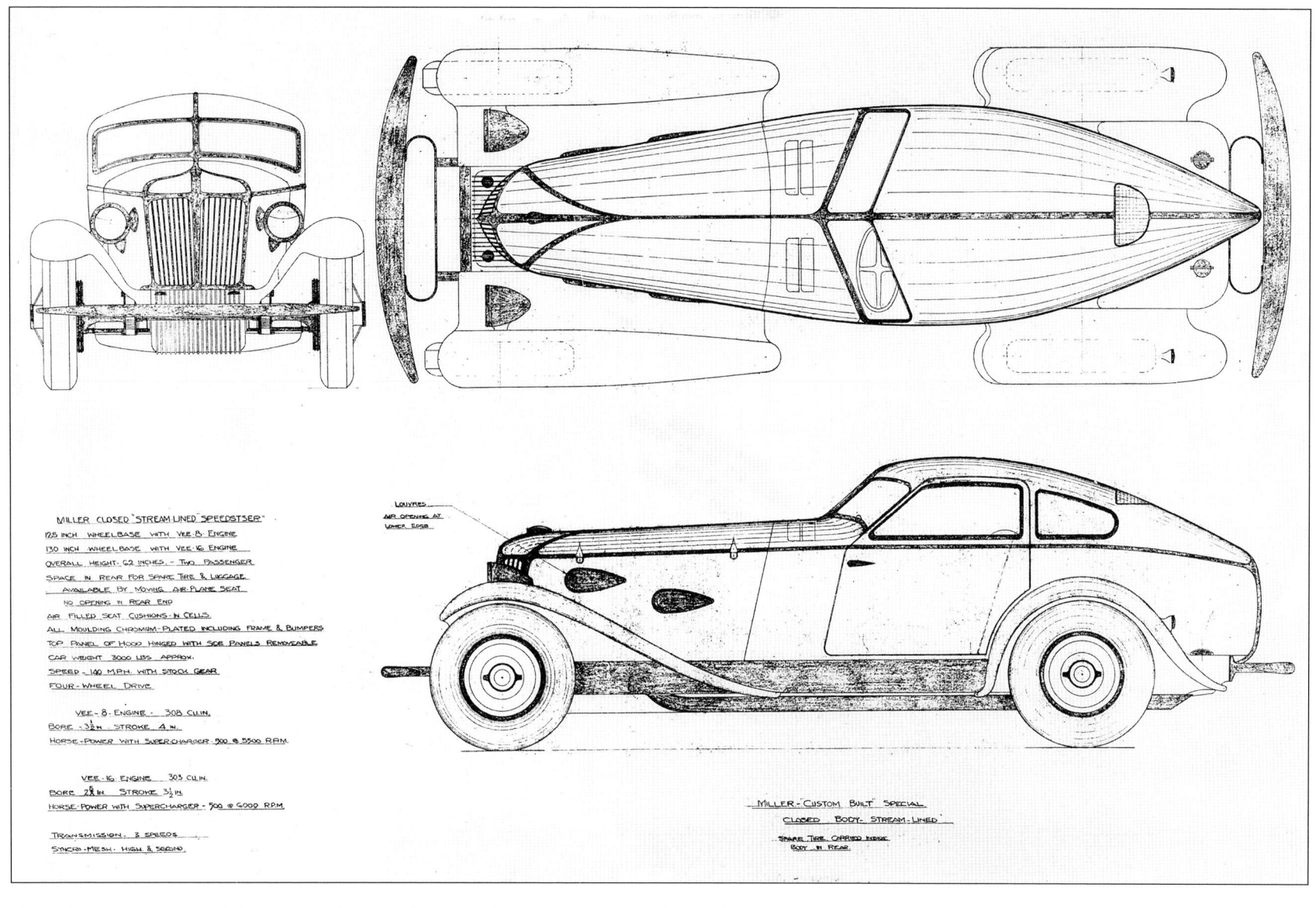

The closed version of the Four Wheel Drive car; this body was quite similar to the Auburn Company's cabin speedster of 1929.

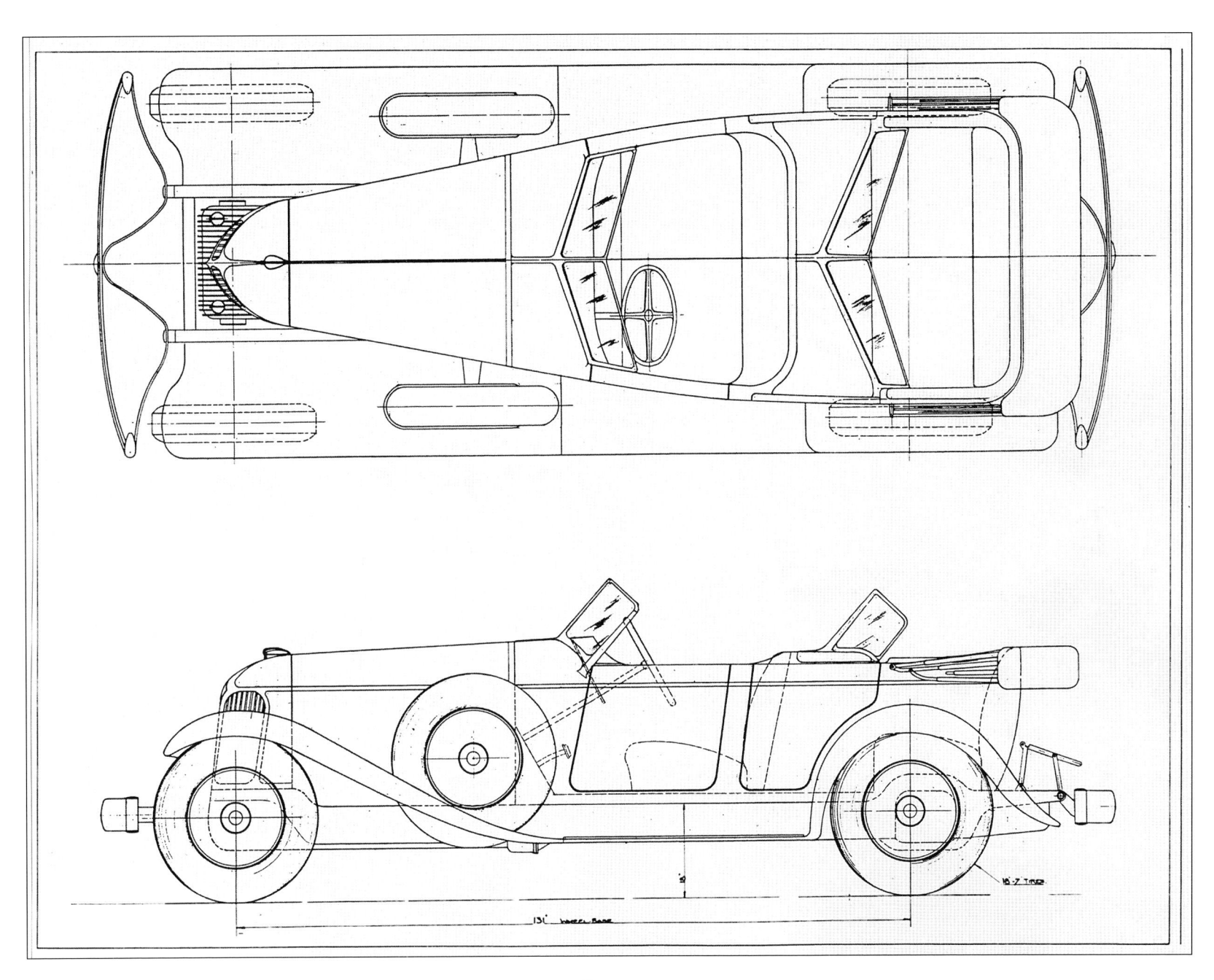

A proposed Miller dual cowl phaeton of 1933; never built.

Miller Passenger car c. 1932 DOHC 4

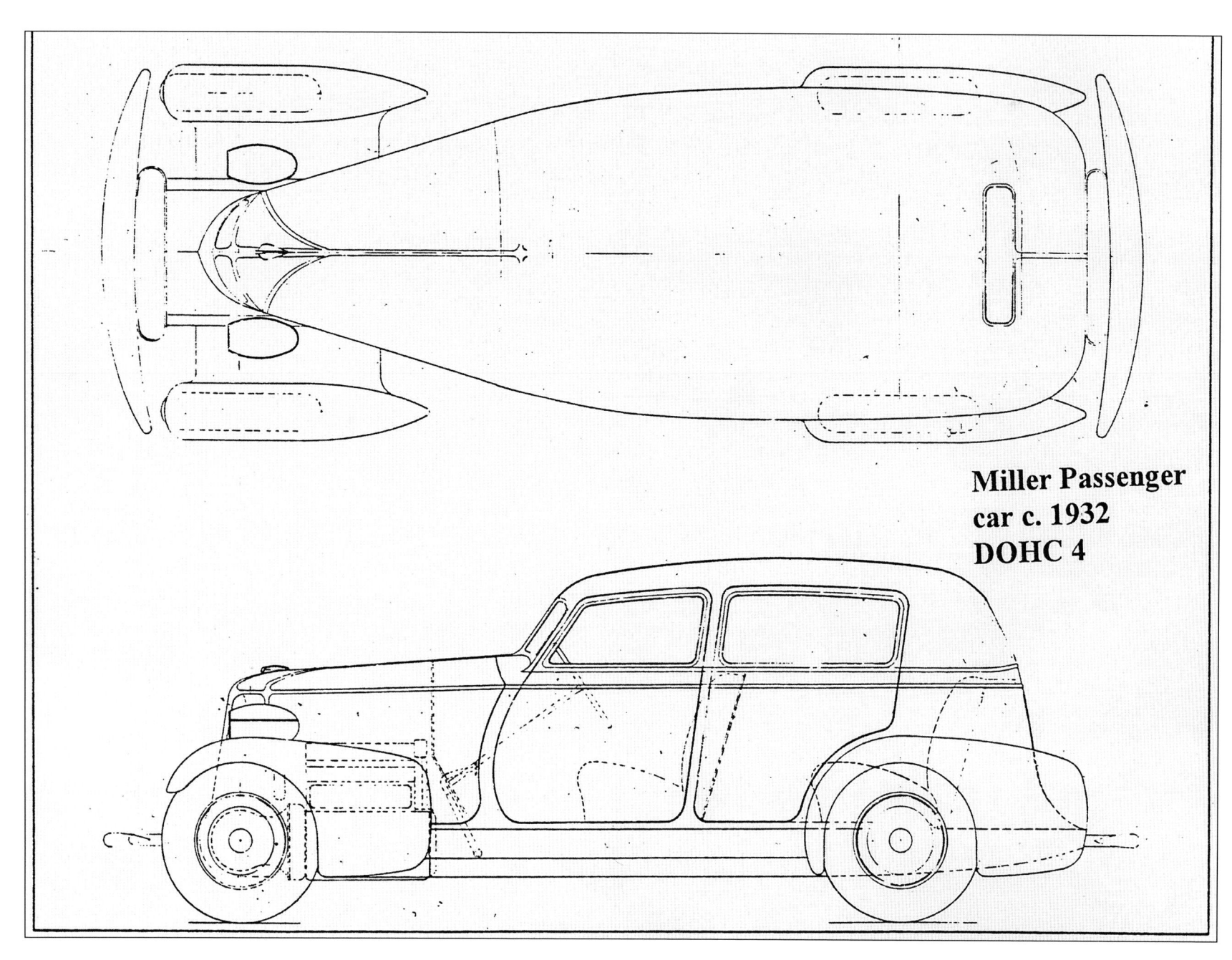

A four-door sedan to be powered by a four-cylinder twin-cam racing type engine.

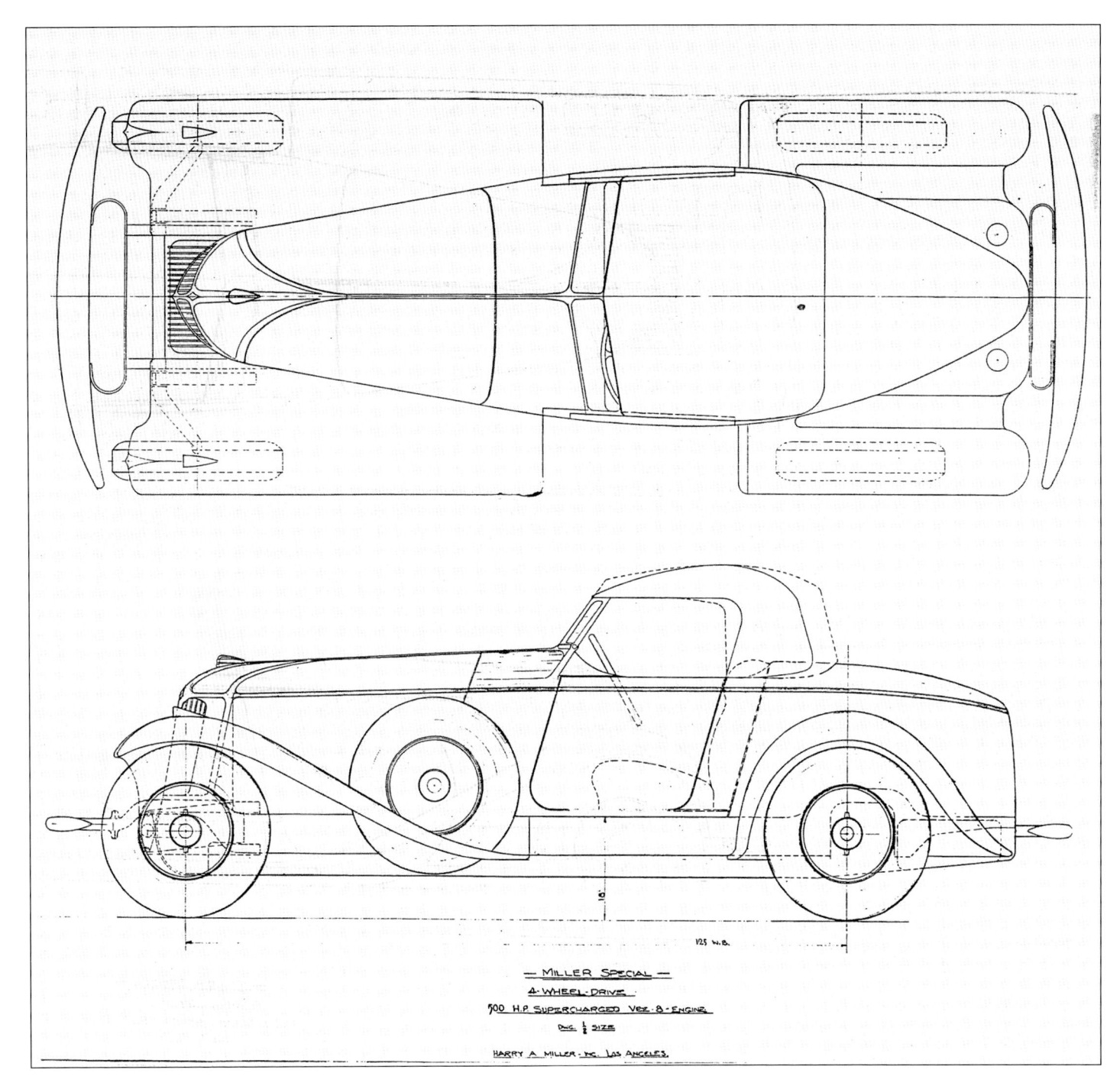

Another Miller four-wheel drive supercharged V-8 open passenger roadster, to have 500 horsepower. What a hot rod it would have made!

95

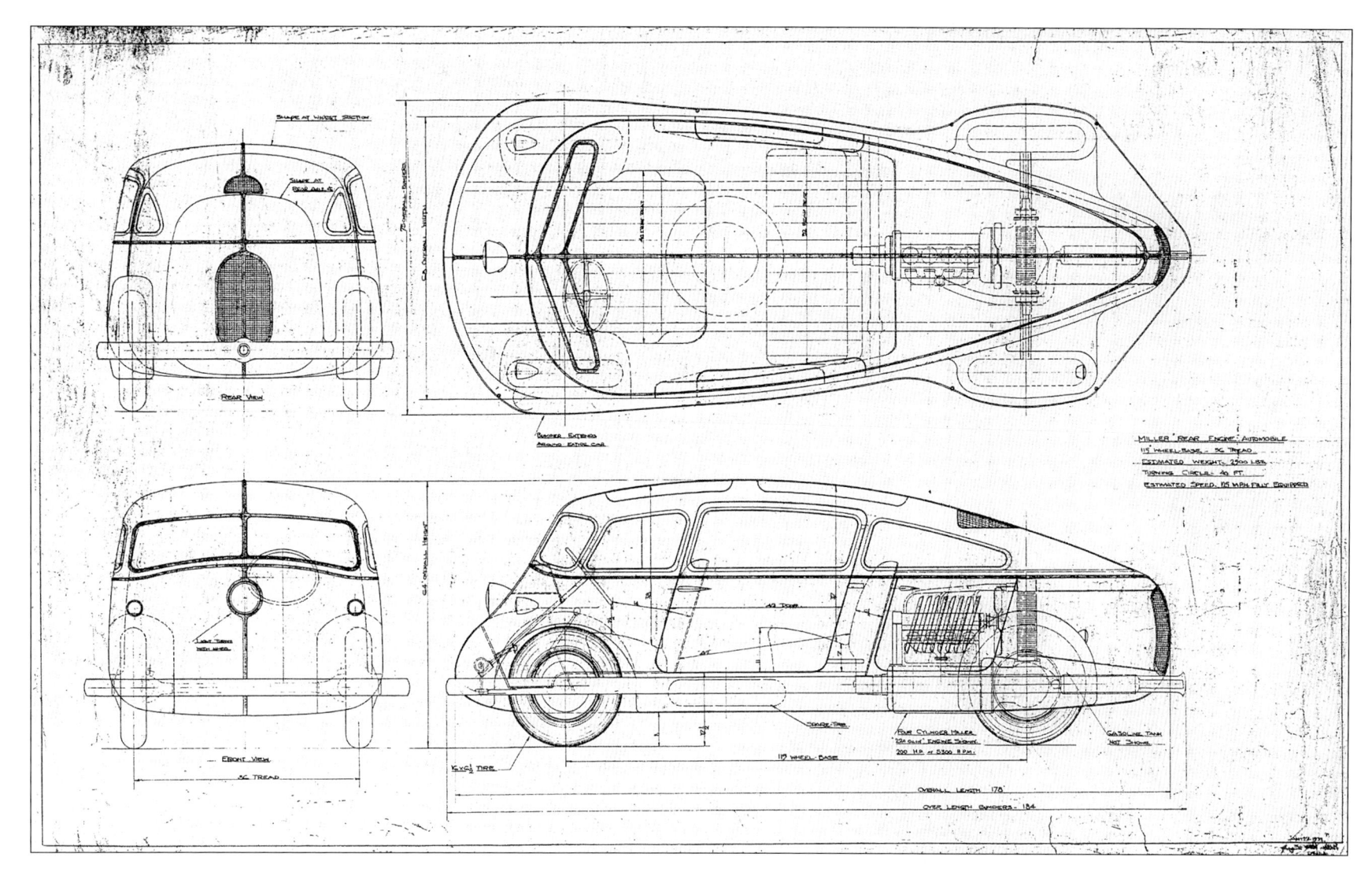

Teardrop shaped rear engine designs were the rage in the late 1920s and early 1930s and Miller appears to have anticipated both Bill Stout's Scarab and Buckminster Fuller's Dymaxion with this version, to be powered by a 220-ci twin-cam four. Note that this one has a single center-mounted headlight that turns with the steering wheel, a feature that would surface, years later, on the Tucker Torpedo of Miller's erstwhile partner, Preston Tucker.

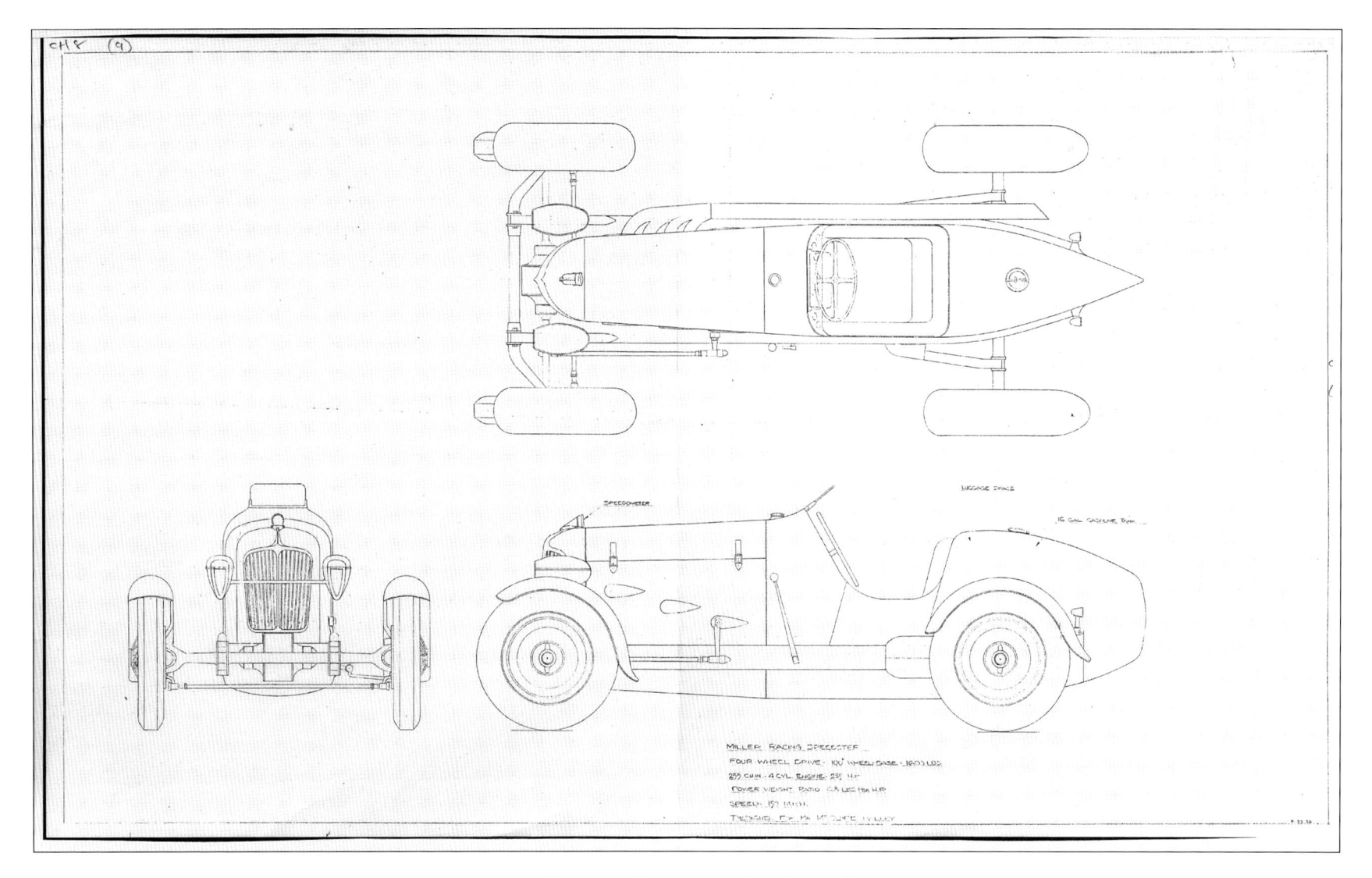

This was another Miller sports car—a race car with headlights and fenders—for customers who never appeared.

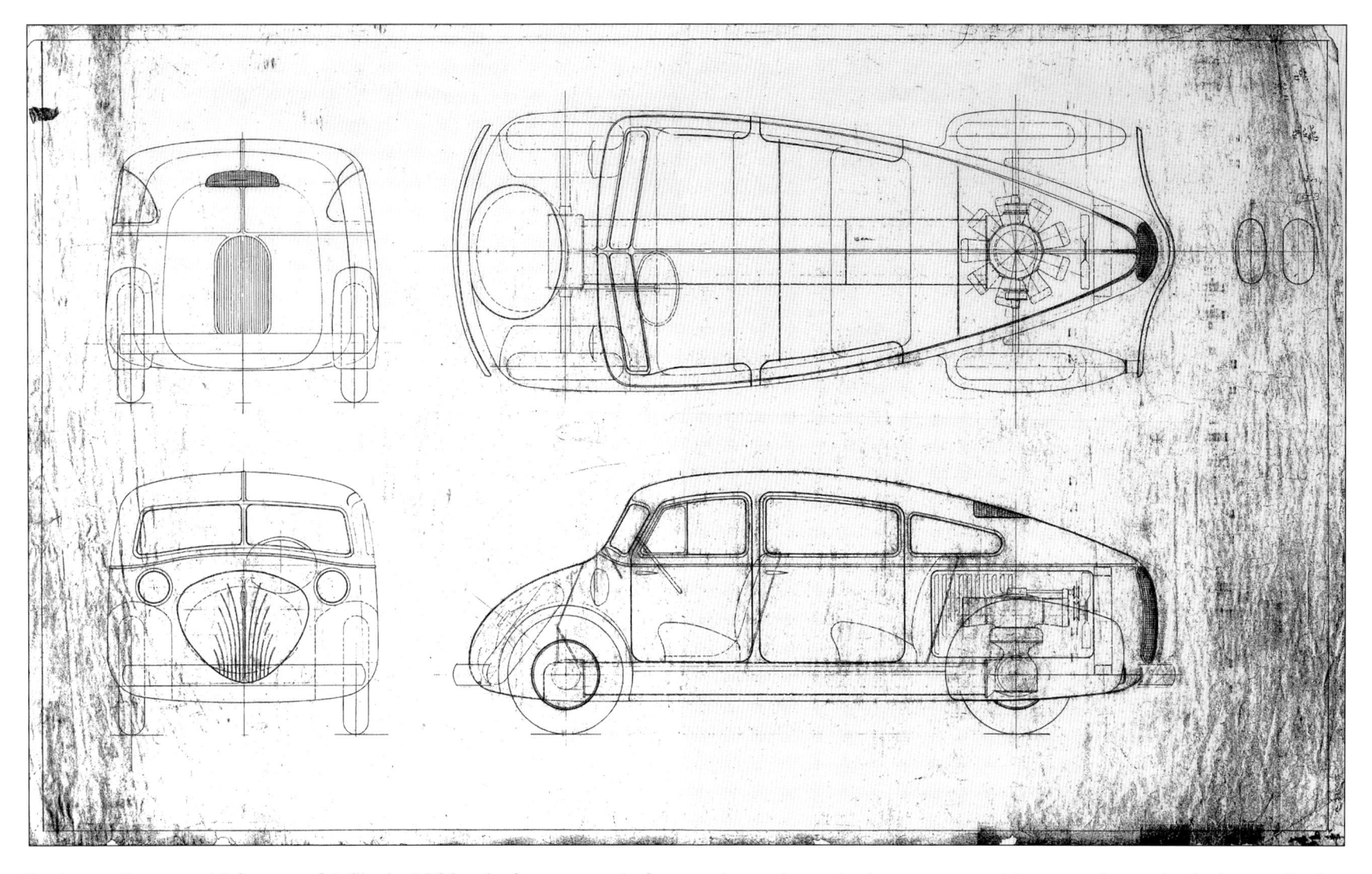

Perhaps the most bizarre of Miller's 1930s designs, a cab-forward road car to be powered by an air-cooled nine-cylinder aircraft type radial engine mounted flat in the rear.

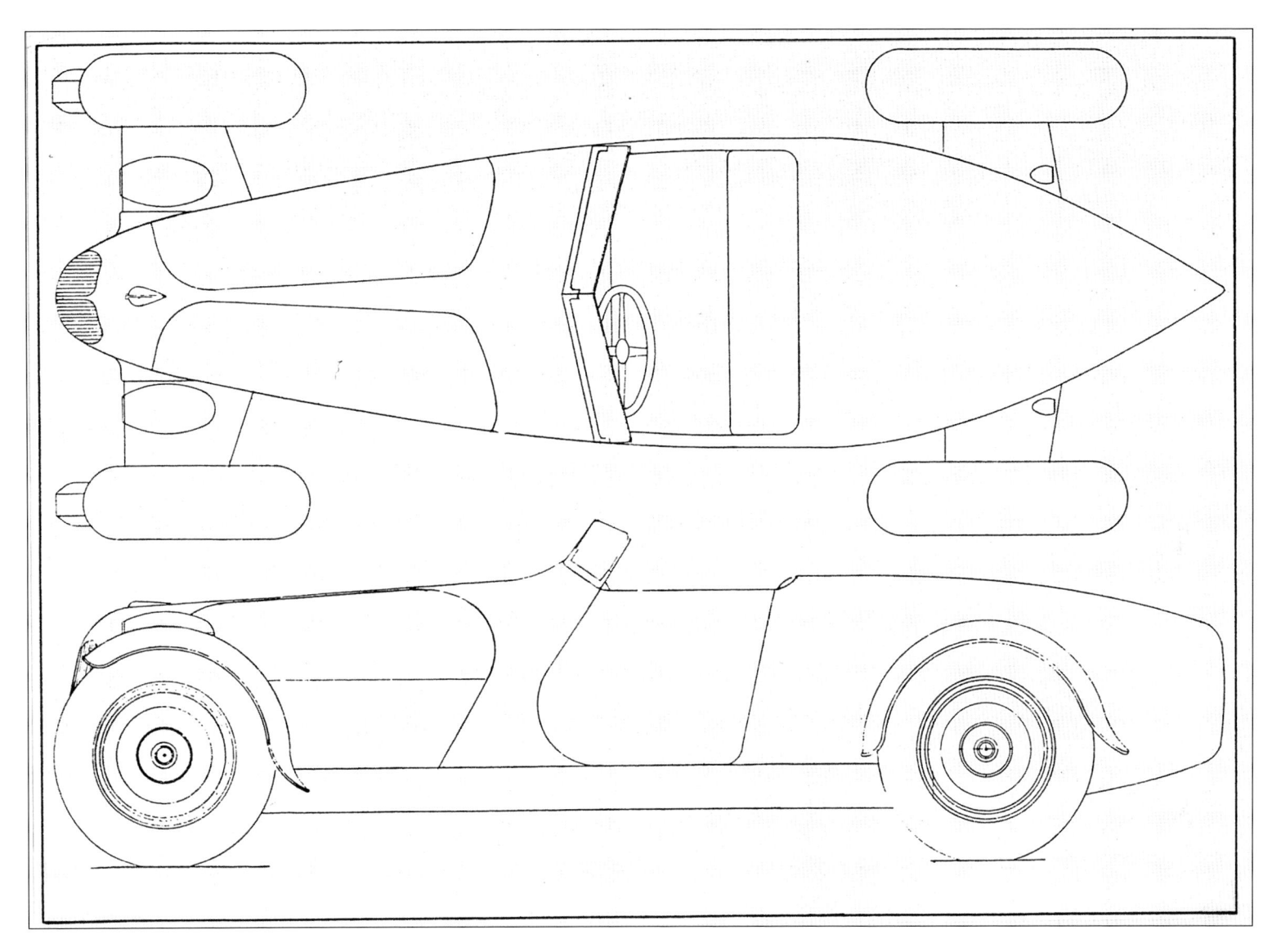

Three versions of this design seem to have anticipated the Plymouth Prowler. All were open cars, "guaranteed" to be capable of 110 miles an hour or more. The first was to be a conventional front engine job, but front-wheel driven. The second was to have been mid-engined with the spare tire and trunk space up front, similar in concept to the Volkswagen.

99

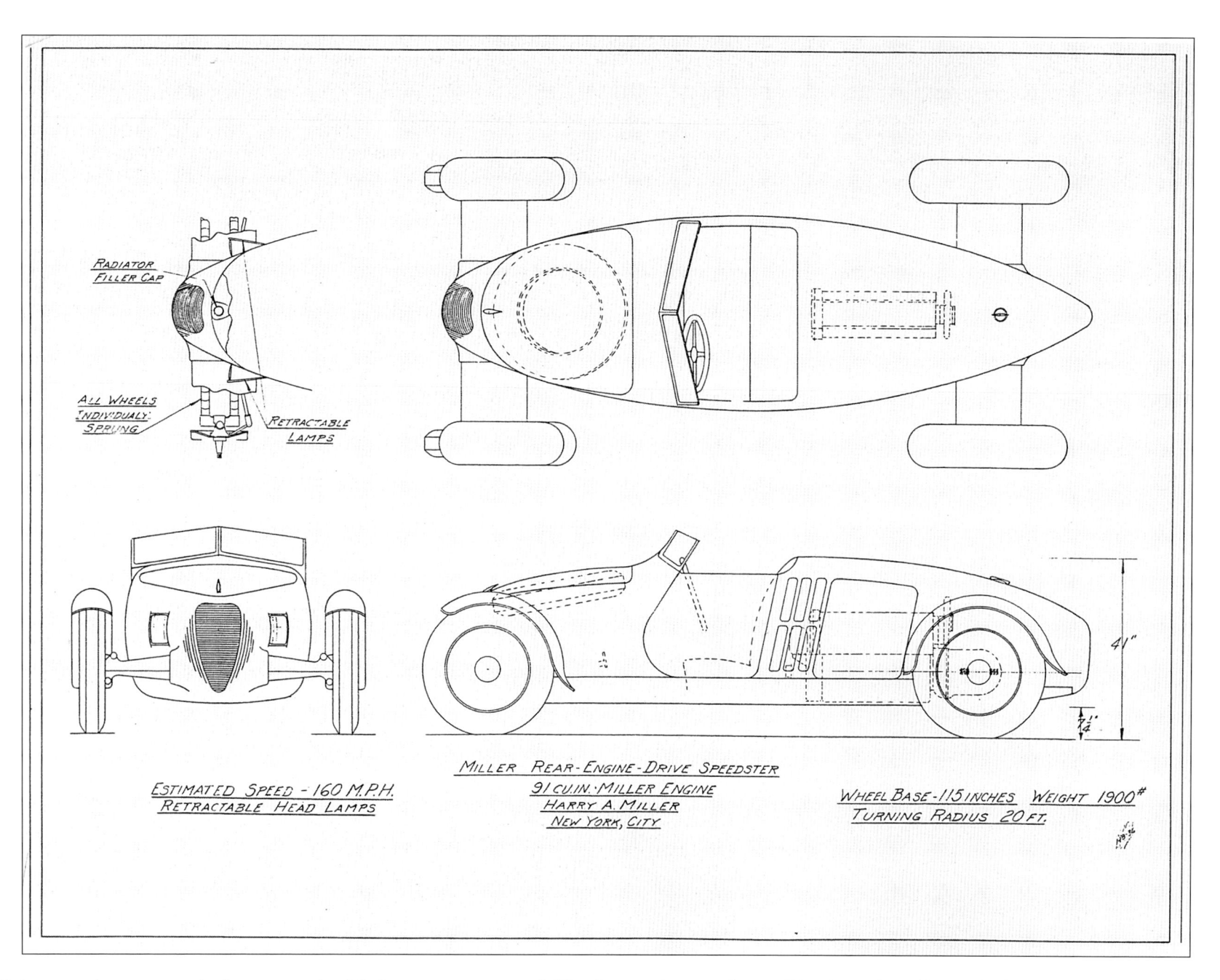

This speedy car was to have carried its two passengers 160 miles an hour using a 91-ci engine. It was to have had side-opening retractable headlights and individual suspension, a design later used on Miller's 1935 Ford Indianapolis cars.

Chapter 9: MILLER-FORDS

Miller, driven from his Los Angeles home by his bankruptcy in 1933, moved to New York and a year later signed a contract to design large aero engines for Floyd L. Brown and David F. Goodnow. He was contacted that winter by Preston Tucker, ever the salesman, who thought that a passenger car engine, dropped into an advanced chassis, might have a chance at Indianapolis. Or at least that one of the big auto companies might pay to find out and provide a nice commission to the man who suggested it. Tucker's idea had more appeal to Miller than the aero engine project so he had his draftsman, R. E. Stevenson, draw some sketches for Tucker, who sold the idea to Edsel Ford.

Tucker told Ford that he and "the famous Harry A. Miller" would produce 10 racing cars for $25,000 plus the provision of 12 engines and other Ford equipment. (He later boosted the price to $75,000.) The cars would become the property of Tucker-Miller Inc. Miller called Stevenson, Walter Leugers, and some others from his old Los Angeles crew to Detroit in early March, prepared to build 10 cars and get them into the race in less than 10 weeks.

A series of six and a half 12-hour-day weeks followed as the race people, aided by "volunteers" from Ford Engineering and subcontractors including Lewis Welch, thrashed to get 10 cars to the Speedway in time for the race. What emerged were rakish, streamlined cars, far more attractive than the usual junk formula cars that had been running at Indy, or even the recycled old Millers that had won the race only five years before. Virtually overnight, Harry produced an automotive tour de force. They were streamlined, even to the airfoil shapes of their independent suspensions and far more aerodynamic than Indianapolis was used to seeing. Power was taken through the front wheels, as in the 1920s Millers. Steering was through an ingenious epicyclic gearbox, mounted via a bracket attached to the engine very near the exhaust manifold.

Their 21-stud Ford V-8 engines were turned around in the chassis and bolted to a front-drive transaxle. The engines were hopped-up with high compression aluminum cylinder heads, racing cams, and put out 150 horsepower. They had four-carburetor intake manifolds and originally used Miller carburetors, but later had Winfields and in one case Stromberg 97s.

At Indy in May the proximity of the steering boxes to the exhausts immediately caused trouble, but it was too late to do much about it except try to duct some air in on them. Only five of the ten were ready when time trials were run and only four made the race. Their drivers were Ted Horn, a second-year rookie who would go on to become national AAA champion; Bob Sall, an eastern champion; Johnny Seymour; and George Bailey.

By the 70-lap mark, three of the cars had gone out with their steering frozen from expansion due to the heat. Horn struggled on until lap 145 when he had to pit and as soon as he rolled to a stop, the steering jammed irretrievably. He was credited with 16th place.

Henry Ford brought the ten cars back to Dearborn and after a legal tiff with Tucker, took possession of them in exchange for forgetting about funds and equipment Tucker had expropriated during their construction. Nearly destroyed by an angry Ford, they lay in storage until Lew Welch was able to buy two of them in 1937. He put a 255 Offenhauser engine in one of the chassis and it finished sixth in 1938 and third in 1939. In 1940, Welch persuaded Ford to bankroll design of a new, supercharged, V-8 engine, known as the "Novi" installed in the 1935 chassis. It placed fourth at Indianapolis in 1941.

The other cars leaked out of Ford's lockup and Marty Keller entered one with an Offy in it at Indy in 1940, with a crew that included a rookie body and chassis man, Frank P. Kurtis. Bus manufacturer Lou Fageol used a pair of spare Miller-Ford transaxles to build a 1946 car that was powered by two separate Offy midget engines. Andy Granatelli put one of their Grancor hot-rodded Mercury V-8s into another that they ran in 1946, 1947, and 1948, until Andy crashed the car in qualifications.

Today, in 2004, all ten still exist. Most have been restored to their 1935 appearance except for the Welch chassis that still carries the original Novi engine.

This shot of one of the 1935 Miller-Fords from overhead shows its clean, streamlined body and the enclosed axles and running gear. The hopped-up Ford V-8 engines were reliable, but somewhat underpowered. Although not likely to win the 500, they might have placed in the top ten had it not been for a flaw in their steering that went uncorrected until race day.

Pit work on qualifying day on a Miller-Ford at Indianapolis in 1935. This is Ted Horn's No. 43 car; it lasted until lap 145.

The front drive of one of the Miller-Fords. The nearly stock Ford V-8 engine was turned around to put the clutch in front. The independently sprung front wheels were suspended from the transmission assembly. Note the use of Ford components such as the tie rod, but the use of four Winfield carburetors.

Cliff Bergere in his Miller-Ford mount. The ten cars were designed and built in the plant on West Lafayette Avenue in Detroit in six and a half weeks.

It is interesting that Lewis Welch chose a Miller-Ford chassis for his Novi V-8 engine in 1941. Henry Ford is reputed to have had a hand or at least some money in the Novi. Ralph Hepburn's fourth place in 1941 was as good as a Novi would ever finish at Indianapolis.

That the basic design was sound was shown by the fact that Miller-Ford chassis continued to run at Indianapolis for another nine years, once the steering problem was identified and corrected. This is Pete Romcevich in the Ford V-8 powered Camco Motors Special, entered by Andy Granatelli. It finished 12th in 1947. The Federal Engineering Special of Henry Banks, powered by an Ottenhauser engine, also used a Miller-Ford chassis in 1947. It finished 24th. Granatelli entered one of these chassis in 1948, but crashed in qualifying.

Chapter 10: THE GULF-MILLERS
THE "CARS FROM MARS"

Griff Borgeson titled an article on the Gulf-Miller racing cars in *Sports Cars Illustrated* magazine "The Cars From Mars" because the rail birds at the Indianapolis Motor Speedway had never seen such wildly radical racing vehicles as Harry Miller brought to the Speedway in May of 1938. The cars had unusual supercharged six-cylinder engines mounted behind the driver, airfoil-shaped external fuel tanks, disc brakes, and independent suspension four-wheel drive. To the Speedway, used to front engines, by rule unsupercharged for seven of the eight prior years, with solid axles and semi-elliptical springs, the latest Millers could well have descended from a Buck Rogers' Martian space ship.

As described earlier, Miller had been working at the American Austin plant in Butler, Pennsylvania, since the Miller–Tucker–Ford debacle of 1935. While there, an old friend, Ira Vail (a driver and later promoter from Syracuse, New York), asked Miller to design and build two Indianapolis cars. Miller brought out drawings Stevenson had done for a proposed four-cylinder aero engine in the dark days in California after Miller's bankruptcy. The planned cars would use a modified version of the Miller-Ford suspension and an early form of disc brakes.

Thirty miles away, in Harmarville, Pennsylvania, the Gulf Oil Co. had just set up a research and development center. Among other things, they investigated the properties of "No-Nox" gasoline and "Gulfpride" lubricating oil. The son of Gulf Chairman James F. Drake heard that Miller was in the vicinity, drove up to Butler, and saw the new cars being built. Soon Vail had been bought out and Miller, Stevenson, and the California crew of Ernie Weill, Emil Deidt, Benny Hill, Eddie Offutt, and Walt Leugers were on their way to the Pittsburgh suburbs.

The four-cylinder engine had the same bore and stroke dimensions as the Miller 255 (by then Offenhauser 255), twin cams but only two valves per cylinder, a split crankcase like the autogiro engine with a detachable cylinder head but without removable bearing webs. Superficially, apart from the aircraft-style dual ignition, it looked a lot like the 255.

Nevertheless, as advanced as the engine seemed to be, Drake wanted the very latest chassis design Miller could conjure up. Harry was happy to oblige. As soon as the entourage got to Harmarville, Miller and Stevenson began work on an even more advanced design: the four-wheel-drive supercharged rear-engined six that was to so startle Indianapolis.

The four put out impressive power on the Gulf dynamometer, even with standard Gulf 80-octane motoring gasoline. The finished cars were taken 300 miles away to the mile dirt track at Langhorne for testing. Immediately, one of Miller's designs, a radiator made of chrome plated copper tubing assembled around the cars' hoods, fractured. The un-finned tubing proved completely inadequate as a radiator, and in its place, conventional radiators were placed on each side of the hoods. (The design of the frame made it difficult to merely mount a radiator in front of the engine in the ordinary way, although that would eventually cure the cooling problem.)

Miller took both the fours and the sixes to Indy in 1938 but none of the cars could qualify. The fours were tested again at Indy in 1939. They were never again entered by Gulf, but were sold to Preston Tucker, who used the engines in mock-ups of landing craft he was proposing to build with Andrew Higgins. They turned up in the hands of an ex-Higgins employee who sold them to a collector in Chicago and then, after many years, went to Chuck Davis, a noted restorer of Miller cars. One of the chassis went through a number of hands before being bought by collector David Willis who, unable to acquire an original engine, installed a power plant made from components of one of the Miller L-510 aero engines, another Miller-Tucker effort from 1940.

The reversal of the supercharger ban at Indianapolis for the 1937 race freed Miller to design the Gulf sixes to use blowers, even though the 80-octane Gulf fuel they had to use was far too low in octane for a supercharged racing engine. The poor quality fuel would prevent the cars from fulfilling their potential. The cars were, of course, grossly over-complex for their day, an example of the loss of the common-sense influence of Goossen and Offenhauser on Miller's flights of technical fantasy during his best years.

The original Gulf sixes used a split crankcase like that of the fours, but returned to the head and block of traditional Miller (and Offenhauser) engines. The first models used Roots type superchargers that were later replaced by centrifugal blowers. The starter was an aircraft type that used a blank shotgun shell for power. Canted to the left by 45 degrees, the engine was a sculpture in cast, ribbed aluminum. Because of its lightness, Miller used tie-down rods to help hold the block to the crankcase, a step taken on the Offenhauser engine by Meyer & Drake in 1951.

In 1939, George Bailey put one of the revamped sixes into the field at Indianapolis, but finished 26th with a broken valve spring. That summer Miller left Gulf's employ for Detroit. Eddie Offutt, who took over the project for Gulf, brought the cars back to Indianapolis in 1940. George Bailey crashed one of them in practice. As Johnny Seymour's car had done in 1939, it burst into flame and Bailey, not as lucky as Seymour, died. The cars were withdrawn. That summer, one of them was run at the Bonneville Salt flats in western Utah, where George Barringer set International Class D records of 153.237 mph for 10 kilometers and 142.799 mph for 500 miles.

The cars were then rebodied by Herman Rigling who put the fuel tanks within the frames. Drivers Barringer and Al Miller managed to qualify both cars in 1941. Barringer's car was destroyed when a fuel fire broke out in its garage; the fire eventually burnt down half a block of the garages in Gasoline Alley. Miller, who started 14th, finished 28th with a sick engine and a broken transmission linkage.

Miller's first design for Gulf was a completion of the four-cylinder car he had begun for Ira Vail. Gloriously innovative, it showed just how much Harry missed Leo Goossen and Fred Offenhauser. Neither of them would have let him build the radiator of chrome-plated tubing wrapped around the nose of the car. It was not only aerodynamically poor, but it was a thermodynamic disaster. The most innovative parts of the car were the disc brakes, probably derived from Bendix brakes first used on the P-6 Curtis fighter aircraft.

Barringer bought the surviving car in 1946 and sold it to Tucker. Entered as the Tucker Torpedo to advertise Tucker's much-hyped passenger car, Barringer qualified it but again dropped out in mid-race. In 1947, Al Miller qualified for the race but could not finish. In 1948, the engine was destroyed in practice. The car is now owned by the Indianapolis Motor Speedway Hall of Fame Museum.

The Gulf Miller four-cylinder engine embodied some of Miller's advanced thinking, with a split crankcase and wet cylinder sleeves, but was otherwise an update of the design he had been working on in 1933 before his bankruptcy.

The chassis for the Gulf-Miller four was adapted from the independent suspension of the Miller-Fords. In this photograph the abortive tubular radiator that did not work has already been replaced by a pair of conventional, finned radiators on each side of the hood. Not exposed at right angles to the airflow, they never worked very well. A radiator mounted in the nose would have worked much better, but that would have required an entire redesign.

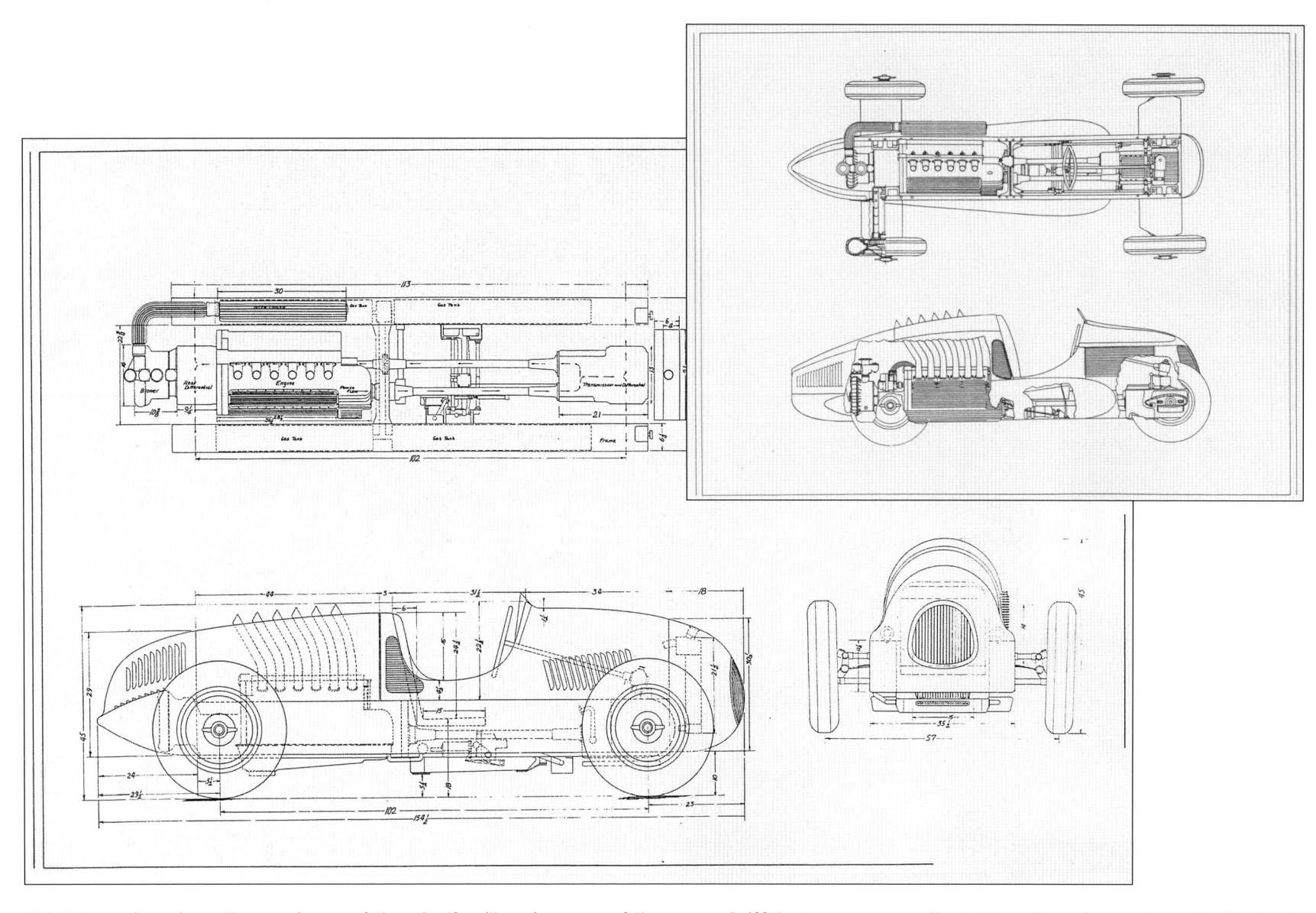

This blueprint gives three views of the Gulf-Miller six, one of the cars Griffith Borgeson called "The Cars from Mars," with their rear-mounted six-cylinder supercharged engines, and four-wheel drive.

George Barringer, who would drive Gulf-Miller cars in 1941 and 1946, with the six-cylinder blown engine.

Harry Miller at Indianapolis in one of his Gulf cars in 1939. In this early version, the exhaust was fed into a horizontal tube. Later, the exhaust stacks were individual, curved vertical stacks.

A rear view of the 1939 Gulf-Miller car with Ralph Hepburn at the wheel. The car never reached qualifying speed that year.

Johnny Seymour's Gulf-Miller car burned after a practice crash at Indianapolis in 1939.

The remaining Gulf-Miller was bought after World War II by Preston Tucker to promote his Tucker passenger car. Driver Al Miller qualified it at 124.848 mph, but went out of the race with a bad magneto.

Chapter 11: LSR, L-510, CHRISTIE COMBAT CAR, MILLER MIDGET

When Harry Miller was in financial trouble in 1931, Preston Tucker, the consummate salesman, acted as Miller's representative to try to sell aviation engines to the U.S. Army Air Corps. The Air Corps did not take the bait, probably because Miller's "3,000-horsepower" engines existed only on paper.

In June 1932, J. Walter Christie asked Miller to cooperate with him on the design of an armored combat car that Christie was designing and eventually sold to the Shah's government in Iran. Christie had built a front-drive racing car for the early Vanderbilt Cup races that Barney Oldfield drove, in fact the car that Oldfield used to turn the first 100-mph lap at the Indianapolis Speedway in a 1916 test session for his sponsor, Firestone Tires.

By the early 1930s, Christie had become a world-renowned designer of both tanks and armored cars, although more honored abroad than at home where the U.S. Army still clung to horse cavalry. Miller sent Leo Goossen east to Elizabeth, New Jersey, to work with Christie on the combat car's design. For Goossen the journey was a perilous one and his existence there hand to mouth. During 1932, Leo's letters to his wife, Vera, tell of surviving on handouts from Christie because Miller was unable to pay his salary. Miller took Leo along to provide technical expertise on visits to wealthy potential customers in New York and Detroit, under luxurious circumstances, while he had to live in YMCAs and eat frugally in cheap diners.

The Christie vehicle was to have been powered by a 785-ci twin-cam flat-12 engine. The specifications called for a speed of 100 mph in the wheeled version and 50 mph on tracks. Unfortunately, despite the fact that George S. Patton was on the evaluation board and supported it, the penurious Army failed to buy the idea. Christie went on to develop and patent a trailing arm independent suspension system for tanks that the British, Polish, and Russian armies bought. The Soviets used it on their T-34 tank, the best armored vehicle of World War II.

Miller built several aviation engines in addition to the Christofferson six and the World War I V-12. In 1927, Max Whittier asked Miller to build an air-cooled aircraft engine for a biplane of his own design. Harry and Leo came up with a 310-ci engine of eight cylinders, horizontally-opposed, with a single cam on each side operating the valves through rocker arms, much the same layout as Continental and Lycoming produce today for light aircraft. Completed but never flown, the design anticipated modern light aircraft engines of this layout by half a dozen years. It produced 130 horsepower at 2,000 rpm—comparable to similar engines built a half-century later.

In October 1935, James Ray of the Pitcairn Autogiro Company in Pennsylvania asked Miller to build a light-weight aero engine to power Harold F. Pitcairn's last autogiro, the AC-35. The autogiro was a predecessor to the helicopter with both a conventional propeller and a large rotor that was usually not powered. Autogiros were capable of very short takeoffs and landing, but could not hover. The overhead rotor acted like a parachute to let the craft down slowly if it lost engine power.

Correspondence and a drawing of the engine show it looked much like the 1934 Offenhauser midget engine but with a split crankcase, like the Gulf-Miller four. The resulting prototype was a 170-pound four-cylinder twin-cam engine of 99.7 ci that produced 164 horsepower at 8,000 rpm. Ray wrote Miller that November saying, "your small engine is very interesting to us and is getting very close to what our engineers have in mind." The AC-135 was to be a "roadable" aircraft, that is, able to travel on the highway as well as in the air and look much like a helicopter. Sponsored in part with a grant from the Federal Bureau of Air Commerce, it was supposed to be "everyman's" aircraft at a price of $700, but it turned out to cost nearer $20,000 in Depression-era dollars. At least one such engine was built and still exists. Pitcairn used another engine in its autogiro.

After he had worn out his welcome at Gulf, Miller felt around desperately for new work. As war clouds gathered more ominously overseas, many companies rushed to offer services to the military. In March 1938, an announcement went out about the formation of the Miller-Bellanca Company, in Montreal, which would manufacture bombing aircraft to be designed by Guiseppe Bellanca and be powered by 1,000-horsepower Miller engines. A year later, an apparently more substantial arrangement appeared on Miller's horizon as discussions ensued with the Bell Aircraft Company for a replacement for the then-unreliable Allison V-1710 engine for the P-39 Aircobra. Miller and Stevenson moved to Indianapolis and began drawings for a V-16 engine for Bell. They also sold plans to the Kermath Company for a 1,725-ci engine for Navy patrol boats. Little of this work came to any useful fruition.

Preston Tucker had seen drawings of the Christie combat car and drew up a variation designed, with a Miller engine, to travel at a supposed 117 mph. The Tucker vehicle had an ingenious revolving turret. Although the army never bought the car, it liked the turrets. In 1940, Tucker was given a contract to manufacture turrets in Ypsilanti, Michigan.

Tucker bought the two Miller four-cylinder engines from Gulf and used them as the basis for an engine he proposed to sell to the Air Corps. He hired Miller to come to Detroit as chief engineer of the Tucker Aircraft Co. Playing on an Air Corps requirement for new, simplified, light, and agile interceptor called, initially, the "Pea Shooter," Tucker drew plans for what the government called the XP-57. For it, Miller designed a pair of eight-cylinder engines, built much like the Gulf fours, and called L-510s.

At least a few prototype L-510 engines were built by Tucker but the XP-57 project was canceled, because it quickly became apparent that heavier, armored, longer-range fighters were needed, not light pea shooters, and because Tucker failed to deliver the prototypes. The money ran short and in August 1940, Miller left Tucker. Miller moved to the tiny shop in Detroit where he worked for the last three years of his life on small wartime subcontracts.

Tucker received a cash infusion in the form of another progress payment from Wright Field that allowed the XP-57 project to limp along until early 1941 and probably to build parts for several L-510 engines. However, after that, the Army pulled the plug. Contract cancellation negotiations dragged on until July 1942. Apparently, Miller got the engines, as they were in his estate when he died in 1943. David Willis acquired some of the blocks and other parts along with one of the Gulf-Miller four chassis and installed half of an L-510 in the race car. Except for minor details, the two engines are much alike.

Several innovative Miller racing cars and engines came off of the drawing boards of Goossen and Stevenson in the early 1930s, few of which were built. One of the most interesting is a December 15, 1932 drawing of a "miniature" race car, virtually identical with the midget racing cars that grew up later in the 1930s. The Miller midget predates by a year in time and at least two years in technology, the midgets that began that sport in California.

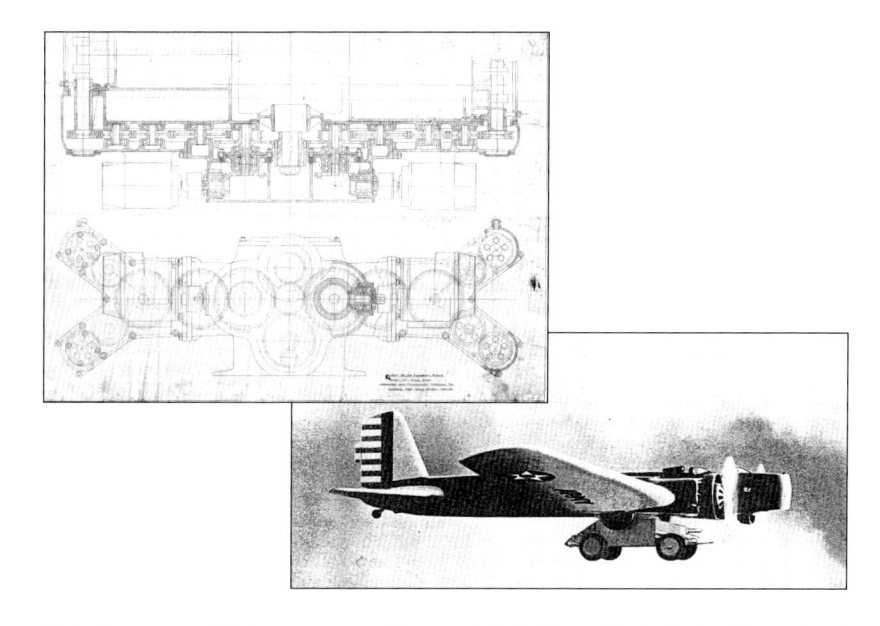

This is an artist's conception of Walter Christie's Combat Car being flown into action in the talons of a B-9 Boeing bomber, a predecessor to the B-17, along with a layout drawing of Miller's 785-ci flat-12-cylinder engine for the Christie Combat Machine.

This is one of the blocks from the Miller L-510 aircraft engine for the XP-57 pursuit plane. It was installed in a Gulf-Miller chassis in an ongoing restoration, as it was the nearest to an original Gulf-Miller four available.

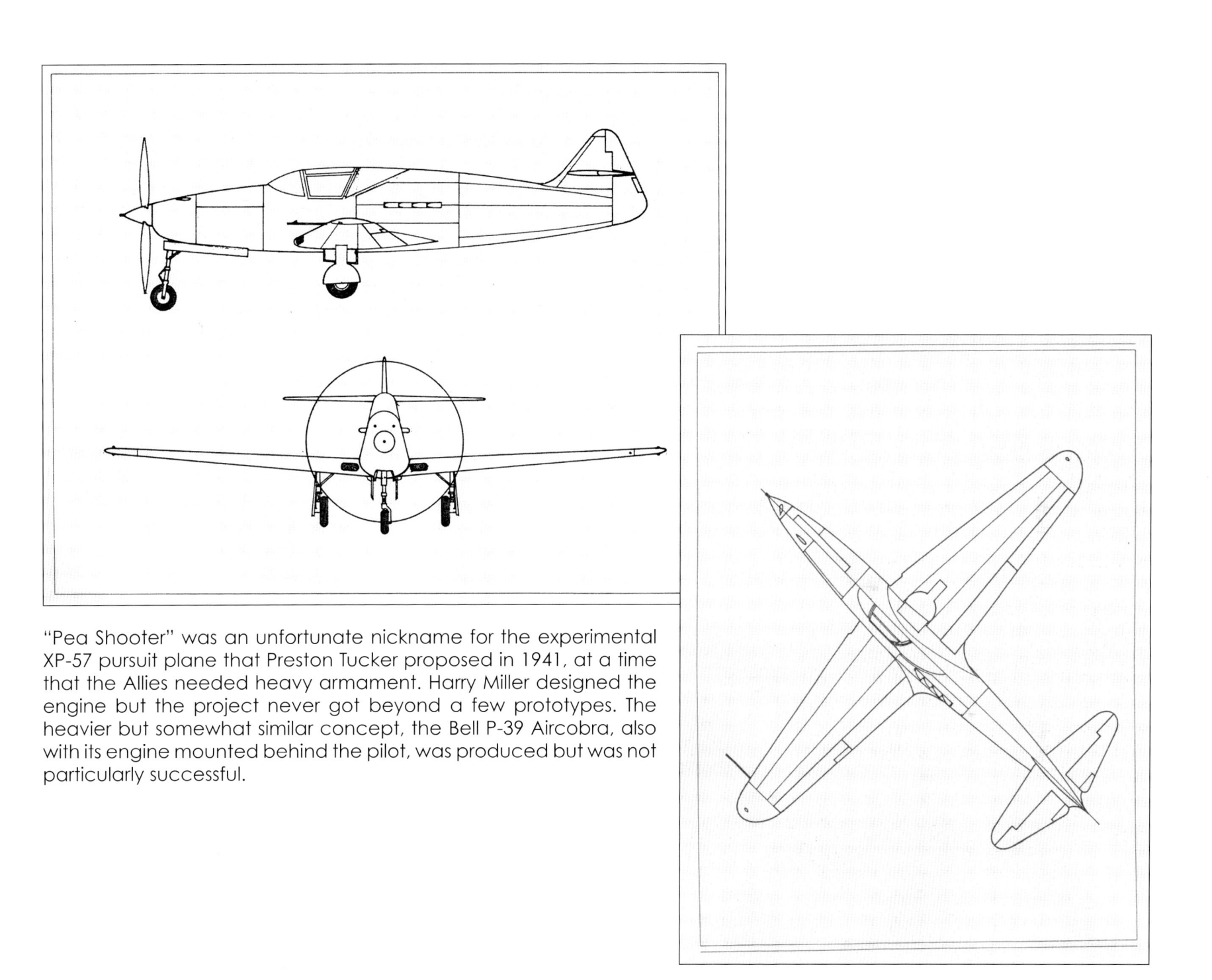

"Pea Shooter" was an unfortunate nickname for the experimental XP-57 pursuit plane that Preston Tucker proposed in 1941, at a time that the Allies needed heavy armament. Harry Miller designed the engine but the project never got beyond a few prototypes. The heavier but somewhat similar concept, the Bell P-39 Aircobra, also with its engine mounted behind the pilot, was produced but was not particularly successful.

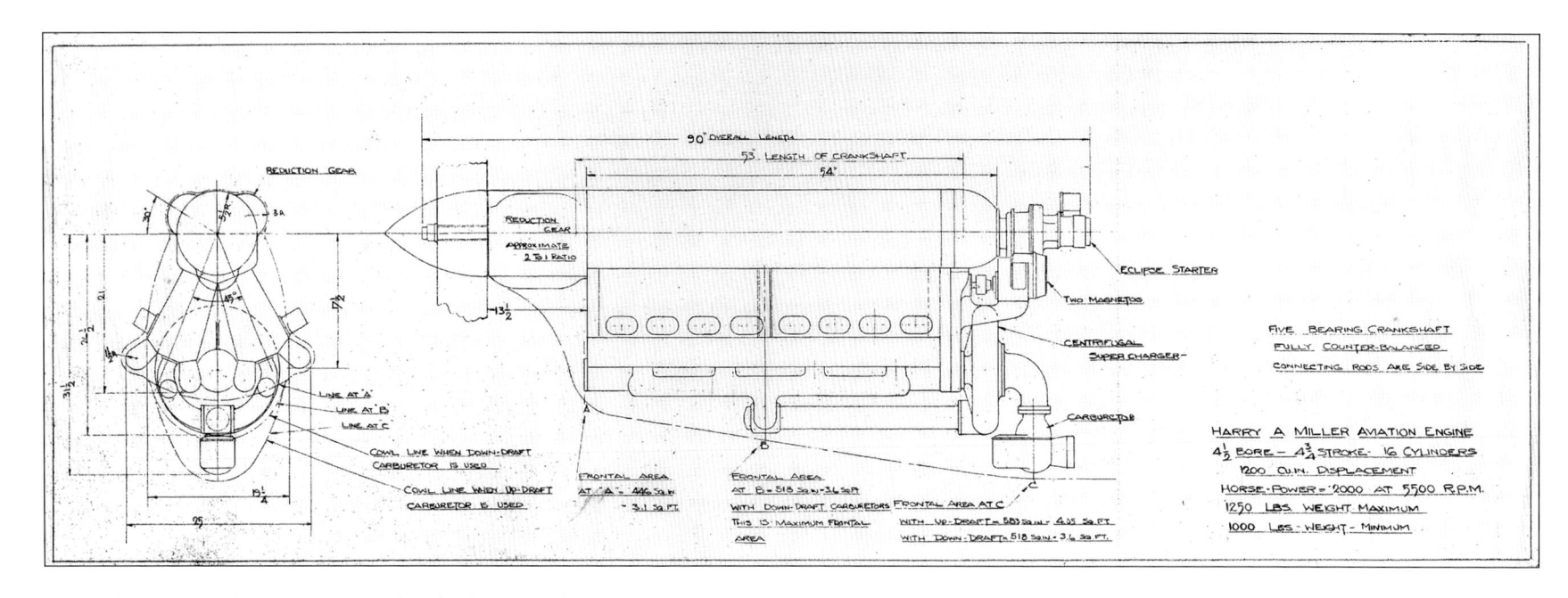

This Miller 16-cylinder, 1,200-ci aviation engine was one that he proposed to the Air Corps just before World War II. He rated it at 2,000 horsepower.

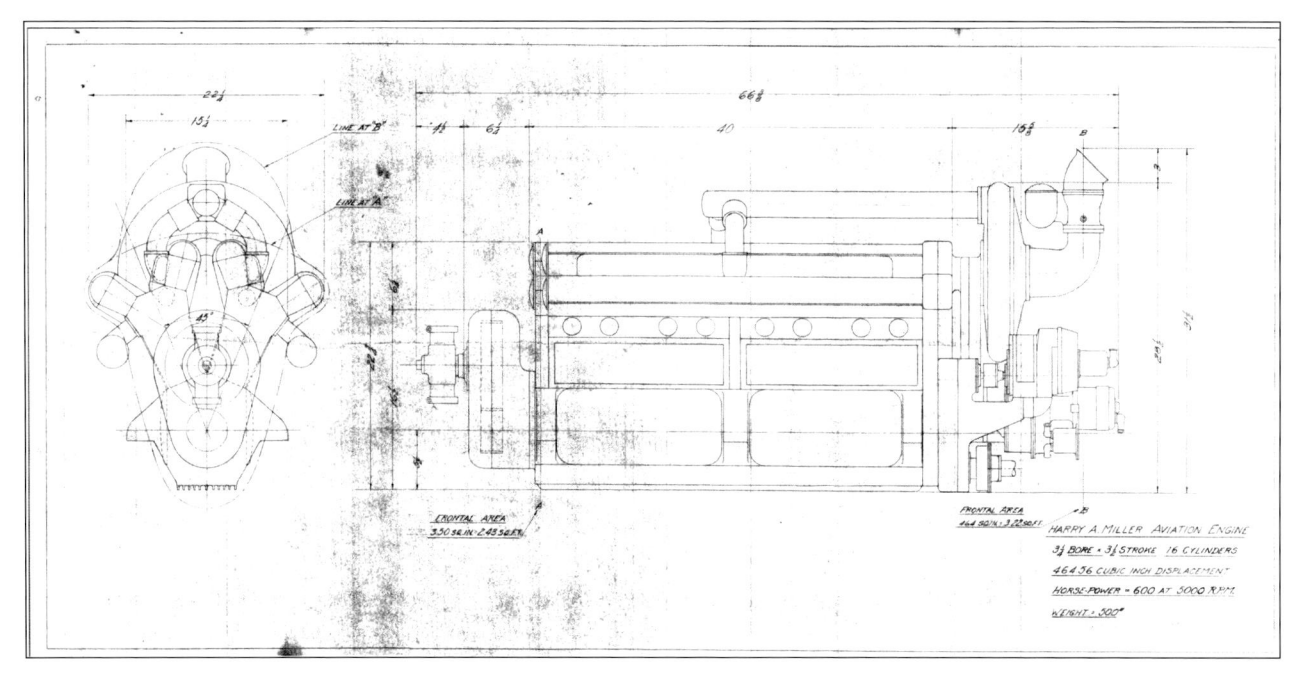

An earlier Miller aviation engine, 464.5 ci, rated at 600 horsepower.

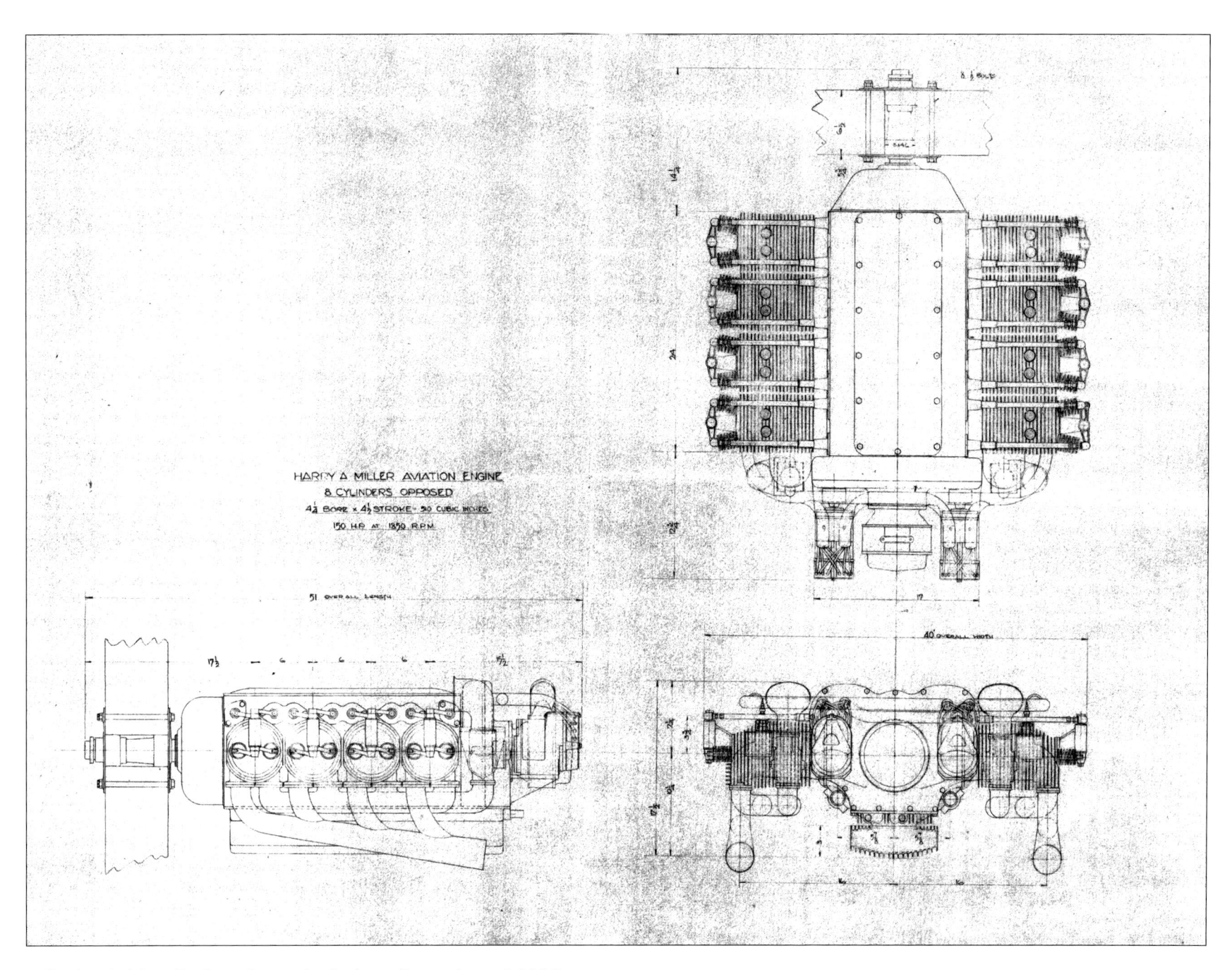

Miller's eight-cylinder air-cooled aircraft engine of 1927.

123

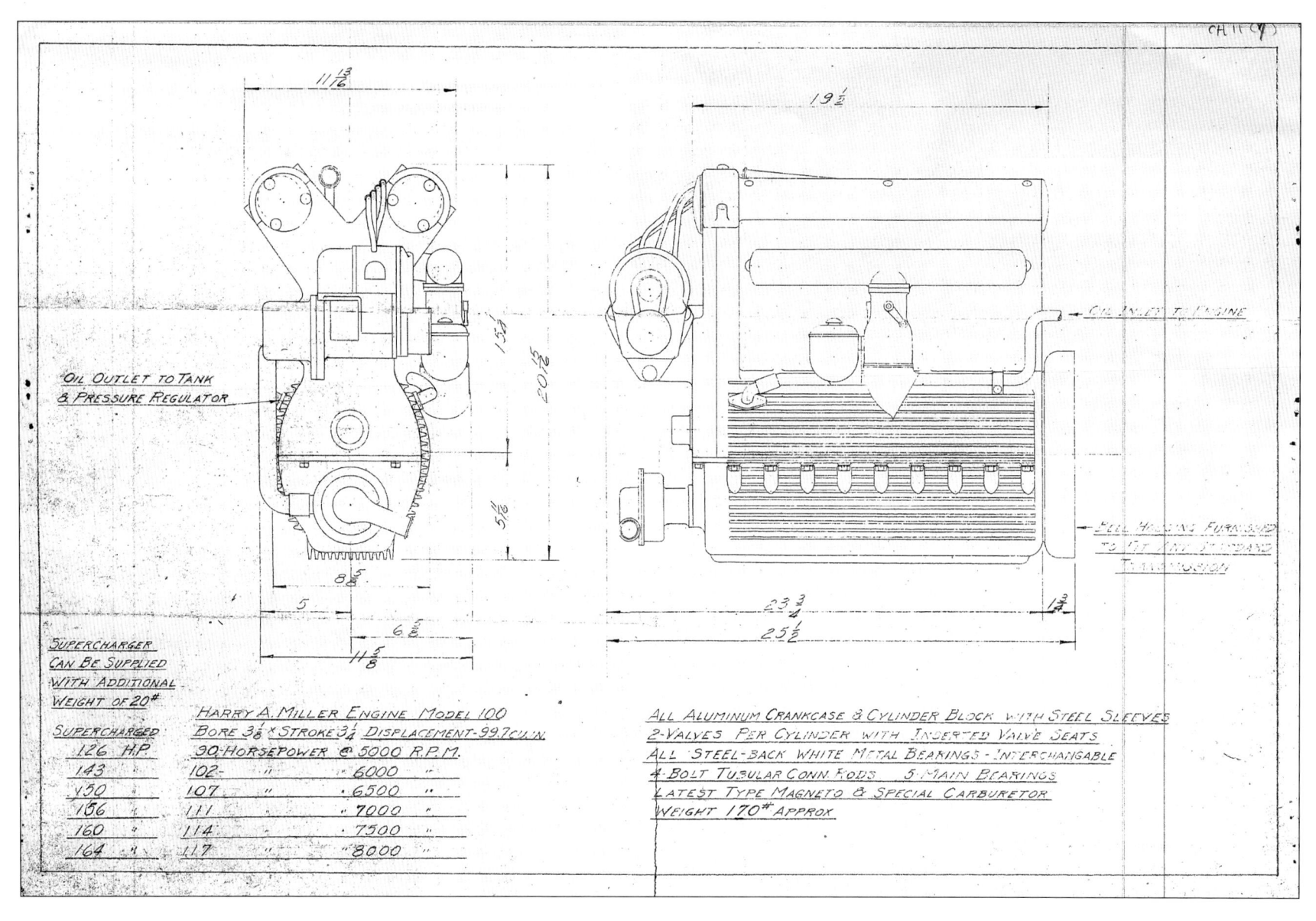

The Miller engine, designed in 1935 for the Pitcairn AC-35 autogiro. Aside from the split crankcase, this engine was quite similar to the 1934 Offenhauser midget engine. (The Offenhauser engine used the original Miller web type main bearing holder.)

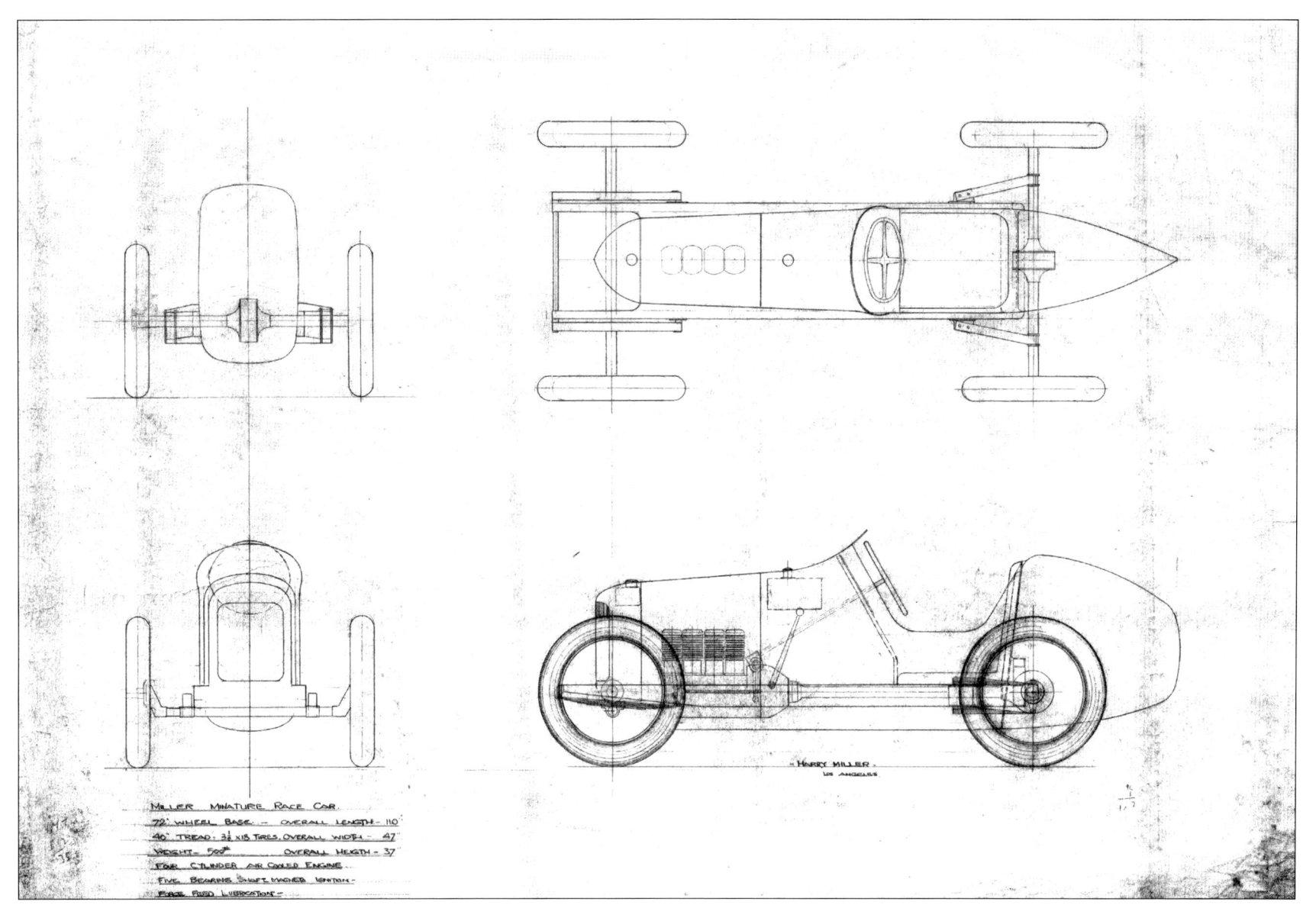

An intriguing might-have-been is this design for a miniature race car about the size of the popular midgets of 1933. This drawing, by Leo Goossen, is dated December 15, 1932.

EPILOGUE

Harry Miller's art typified the decade of the Roaring Twenties—the years of fast women, fast cars, and bootleg liquor. His cars were exquisite and highly-strung, tiny but powerful jewels, more art than science. He was defeated by the Junk Formula at Indianapolis and by the Depression of the 1930s, an era that was not as forgiving as had been the 1920s.

His mind leapt ahead of his technology to ideas both far advanced and impractical, such as the road car with a flat-mounted radial air-cooled engine in the rear and a 100-mph "combat car."

Miller died in 1943, in the midst of World War II. There had been no racing at Indianapolis for two years and it would be three more years until championship racing resumed in the United States. When Miller attempted to develop engines for the Air Corps, the engineers at Wright field asked him for design studies showing stress and vibration analyses that a room full of graduate engineers would have taken a month to develop. As Mark Dees concluded, Miller, the cut and try empiricist, working without the expertise of a Leo Goossen and the machine shop common sense of a Fred Offenhauser, was no longer in the mainstream of engine design. It had become a different world. Even the Air Corps' Wright Field made errors in the design of the Allison V-1710 engine that became obvious under wartime stress. The supercharged V-12 engine that would power the P-51 Mustang to combat glory was not the Allison but the British Rolls-Royce Merlin.

Nevertheless, Harry in his prime was a grand and great artist in metal. His cars and engines still roar at gatherings of those who appreciate the whine of a Miller supercharger, the gleam of polished paint, and the glint of plated metal. His marvelous mechanical designs live on in a new century.

More Great Titles From Iconografix

All Iconografix books are available from direct mail specialty book dealers and bookstores worldwide, or can be ordered from the publisher. For book trade and distribution information or to add your name to our mailing list and receive a **FREE CATALOG** contact:

Iconografix, Inc.
PO Box 446, Dept BK
Hudson, WI, 54016

Telephone: (715) 381-9755,
(800) 289-3504 (USA),
Fax: (715) 381-9756

*This product is sold under license from Mack Trucks, Inc. Mack is a registered Trademark of Mack Trucks, Inc. All rights reserved.

RACING

Chaparral Can-Am Racing Cars from Texas, Ludvigsen Library Series	ISBN 1-58388-066-6
Cunningham Sports Cars, Ludvigsen Library Series	ISBN 1-58388-109-3
Drag Racing Funny Cars of the 1960s Photo Archive	ISBN 1-58388-097-6
Drag Racing Funny Cars of the 1970s Photo Archive	ISBN 1-58388-068-2
El Mirage Impressions: Dry Lakes Land Speed Racing	ISBN 1-58388-059-3
Indy Cars of the 1940s, Ludvigsen Library Series	ISBN 1-58388-117-4
Indy Cars of the 1950s, Ludvigsen Library Series	ISBN 1-58388-018-6
Indy Cars of the 1960s, Ludvigsen Library Series	ISBN 1-58388-052-6
Indy Cars of the 1970s, Ludvigsen Library Series	ISBN 1-58388-098-4
Indianapolis Racing Cars of Frank Kurtis 1941-1963 Photo Archive	ISBN 1-58388-026-7
Juan Manuel Fangio World Champion Driver Series Photo Album	ISBN 1-58388-008-9
Lost Race Tracks Treasures of Automobile Racing	ISBN 1-58388-084-4
Mario Andretti World Champion Driver Series Photo Album	ISBN 1-58388-009-7
Marvelous Mechanical Designs of Harry A. Miller	ISBN 1-58388-123-9
Mercedes-Benz 300SL Racing Cars 1952-1953, Ludvigsen Library Series	ISBN 1-58388-067-4
Mercedes-Benz 300SLR: The Fabulous 1955 World-Champion Sports-Racer	ISBN 1-882256-122-0
MG Record-Breakers from Abingdon Photo Archive	ISBN 1-58388-116-6
Novi V-8 Indy Cars 1941-1965, Ludvigsen Library Series	ISBN 1-58388-037-2
Porsche Spyders Type 550 1953-1956, Ludvigsen Library Series	ISBN 1-58388-092-5
Sebring 12-Hour Race 1970 Photo Archive	ISBN 1-882256-20-4
Top Fuel Dragsters of the 1970s Photo Archive	ISBN 1-58388-128-X
Vanderbilt Cup Race 1936 & 1937 Photo Archive	ISBN 1-882256-66-2

AUTOMOTIVE

AMC Cars 1954-1987: An Illustrated History	ISBN 1-58388-112-3
AMC Performance Cars 1951-1983 Photo Archive	ISBN 1-58388-127-1
AMX Photo Archive: From Concept to Reality	ISBN 1-58388-062-3
Auburn Automobiles 1900-1936 Photo Archive	ISBN 1-58388-093-3
Camaro 1967-2000 Photo Archive	ISBN 1-58388-032-1
Checker Cab Co. Photo History	ISBN 1-58388-100-X
Chevrolet Corvair Photo History	ISBN 1-58388-118-2
Chevrolet Station Wagons 1946-1966 Photo Archive	ISBN 1-58388-069-0
Classic American Limousines 1955-2000 Photo Archive	ISBN 1-58388-041-0
Cord Automobiles L-29 & 810/812 Photo Archive	ISBN 1-58388-102-6
Corvair by Chevrolet Experimental & Production Cars 1957-1969, Ludvigsen Library Series	ISBN 1-58388-058-5
Corvette The Exotic Experimental Cars, Ludvigsen Library Series	ISBN 1-58388-017-8
Corvette Prototypes & Show Cars Photo Album	ISBN 1-882256-77-8
Early Ford V-8s 1932-1942 Photo Album	ISBN 1-882256-97-2
Ferrari- The Factory Maranello's Secrets 1950-1975, Ludvigsen Library Series	ISBN 1-58388-085-2
Ford Postwar Flatheads 1946-1953 Photo Archive	ISBN 1-58388-080-1
Ford Station Wagons 1929-1991 Photo History	ISBN 1-58388-103-4
Hudson Automobiles 1934-1957 Photo Archive	ISBN 1-58388-110-7
Imperial 1955-1963 Photo Archive	ISBN 1-882256-22-0
Imperial 1964-1968 Photo Archive	ISBN 1-882256-23-9
Javelin Photo Archive: From Concept to Reality	ISBN 1-58388-071-2
Lincoln Motor Cars 1920-1942 Photo Archive	ISBN 1-882256-57-3
Lincoln Motor Cars 1946-1960 Photo Archive	ISBN 1-882256-58-1
Nash 1936-1957 Photo Archive	ISBN 1-58388-086-0
Packard Motor Cars 1935-1942 Photo Archive	ISBN 1-882256-44-1
Packard Motor Cars 1946-1958 Photo Archive	ISBN 1-882256-45-X
Pontiac Dream Cars, Show Cars & Prototypes 1928-1998 Photo Album	ISBN 1-882256-93-X
Pontiac Firebird Trans-Am 1969-1999 Photo Album	ISBN 1-882256-95-6
Pontiac Firebird 1967-2000 Photo History	ISBN 1-58388-028-3
Rambler 1950-1969 Photo Archive	ISBN 1-58388-078-X
Stretch Limousines 1928-2001 Photo Archive	ISBN 1-58388-070-4
Studebaker 1933-1942 Photo Archive	ISBN 1-882256-24-7
Studebaker Hawk 1956-1964 Photo Archive	ISBN 1-58388-094-1
Studebaker Lark 1959-1966 Photo Archive	ISBN 1-58388-107-7
Ultimate Corvette Trivia Challenge	ISBN 1-58388-035-6

RECREATIONAL VEHICLES

RVs & Campers 1900-2000: An Illustrated History	ISBN 1-58388-064-X
Ski-Doo Racing Sleds 1960-2003 Photo Archive	ISBN 1-58388-105-0

TRUCKS

Autocar Trucks 1899-1950 Photo Archive	ISBN 1-58388-115-8
Autocar Trucks 1950-1987 Photo Archive	ISBN 1-58388-072-0
Beverage Trucks 1910-1975 Photo Archive	ISBN 1-882256-60-3
Brockway Trucks 1948-1961 Photo Archive*	ISBN 1-882256-55-7
Chevrolet El Camino Photo History Incl. GMC Sprint & Caballero	ISBN 1-58388-044-5
Circus and Carnival Trucks 1923-2000 Photo Archive	ISBN 1-58388-048-8
Dodge B-Series Trucks Restorer's & Collector's Reference Guide and History	ISBN 1-58388-087-9
Dodge Pickups 1939-1978 Photo Album	ISBN 1-882256-82-4
Dodge Power Wagons 1940-1980 Photo Archive	ISBN 1-882256-89-1
Dodge Power Wagon Photo History	ISBN 1-58388-019-4
Dodge Ram Trucks 1994-2001 Photo History	ISBN 1-58388-051-8
Dodge Trucks 1929-1947 Photo Archive	ISBN 1-882256-36-0
Dodge Trucks 1948-1960 Photo Archive	ISBN 1-882256-37-9
Ford 4x4s 1935-1990 Photo History	ISBN 1-58388-079-8
Ford Heavy-Duty Trucks 1948-1998 Photo History	ISBN 1-58388-043-7
Ford Ranchero 1957-1979 Photo History	ISBN 1-58388-126-3
Freightliner Trucks 1937-1981 Photo Archive	ISBN 1-58388-090-9
GMC Heavy-Duty Trucks 1927-1987	ISBN 1-58388-125-5
Jeep 1941-2000 Photo Archive	ISBN 1-58388-021-6
Jeep Prototypes & Concept Vehicles Photo Archive	ISBN 1-58388-033-X
Mack Model AB Photo Archive*	ISBN 1-882256-18-2
Mack AP Super-Duty Trucks 1926-1938 Photo Archive*	ISBN 1-882256-54-9
Mack Model B 1953-1966 Volume 2 Photo Archive*	ISBN 1-882256-34-4
Mack EB-EC-ED-EE-EF-EG-DE 1936-1951 Photo Archive*	ISBN 1-882256-29-8
Mack EH-EJ-EM-EQ-ER-ES 1936-1950 Photo Archive*	ISBN 1-882256-39-5
Mack FC-FCSW-NW 1936-1947 Photo Archive*	ISBN 1-882256-28-X
Mack FG-FH-FJ-FK-FN-FP-FT-FW 1937-1950 Photo Archive*	ISBN 1-882256-35-2
Mack LF-LH-LJ-LM-LT 1940-1956 Photo Archive*	ISBN 1-58388-38-7
Mack Trucks Photo Gallery*	ISBN 1-882256-88-3
New Car Carriers 1910-1998 Photo Album	ISBN 1-882256-98-0
Plymouth Commercial Vehicles Photo Archive	ISBN 1-58388-004-6
Refuse Trucks Photo Archive	ISBN 1-58388-042-9
Studebaker Trucks 1927-1940 Photo Archive	ISBN 1-882256-40-9
White Trucks 1900-1937 Photo Archive	ISBN 1-882256-80-8

EMERGENCY VEHICLES

The American Ambulance 1900-2002: An Illustrated History	ISBN 1-58388-081-X
American Fire Apparatus Co. 1922-1993 Photo Archive	ISBN 1-58388-131-X
American Funeral Vehicles 1883-2003 Illustrated History	ISBN 1-58388-104-2
American LaFrance 700 Series 1945-1952 Photo Archive	ISBN 1-882256-90-5
American LaFrance 700 Series 1945-1952 Photo Archive Volume 2	ISBN 1-58388-025-9
American LaFrance 700 & 800 Series 1953-1958 Photo Archive	ISBN 1-882256-91-3
American LaFrance 900 Series 1958-1964 Photo Archive	ISBN 1-58388-002-X
Classic Seagrave 1935-1951 Photo Archive	ISBN 1-58388-034-8
Crown Firecoach 1951-1985 Photo Archive	ISBN 1-58388-047-X
Encyclopedia of Canadian Fire Apparatus	ISBN 1-58388-119-0
Fire Chief Cars 1900-1997 Photo Album	ISBN 1-882256-87-5
Hahn Fire Apparatus 1923-1990 Photo Archive	ISBN 1-58388-077-1
Heavy Rescue Trucks 1931-2000 Photo Gallery	ISBN 1-58388-045-3
Imperial Fire Apparatus 1969-1976 Photo Archive	ISBN 1-58388-091-7
Industrial and Private Fire Apparatus 1925-2001 Photo Archive	ISBN 1-58388-049-6
Mack Model C Fire Trucks 1957-1967 Photo Archive*	ISBN 1-58388-014-3
Mack Model L Fire Trucks 1940-1954 Photo Archive*	ISBN 1-882256-86-7
Maxim Fire Apparatus 1914-1989 Photo Archive	ISBN 1-58388-050-X
Maxim Fire Apparatus Photo History	ISBN 1-58388-111-5
Navy & Marine Corps Fire Apparatus 1836-2000 Photo Gallery	ISBN 1-58388-031-3
Pierre Thibault Ltd. Fire Apparatus 1918-1990 Photo Archive	ISBN 1-58388-074-7
Pirsch Fire Apparatus 1890-1991 Photo Archive	ISBN 1-58388-082-8
Police Cars: Restoring, Collecting & Showing America's Finest Sedans	ISBN 1-58388-046-1
Saulsbury Fire Rescue Apparatus 1956-2003 Photo Archive	ISBN 1-58388-106-9
Seagrave 70th Anniversary Series Photo Archive	ISBN 1-58388-001-1
Seagrave Fire Apparatus 1959-2004 Photo Archive	ISBN 1-58388-132-8
TASC Fire Apparatus 1946-1985 Photo Archive	ISBN 1-58388-065-8
Volunteer & Rural Fire Apparatus Photo Gallery	ISBN 1-58388-005-4
W.S. Darley & Co. Fire Apparatus 1908-2000 Photo Archive	ISBN 1-58388-061-5
Wildland Fire Apparatus 1940-2001 Photo Gallery	ISBN 1-58388-056-9
Young Fire Equipment 1932-1991 Photo Archive	ISBN 1-58388-015-1

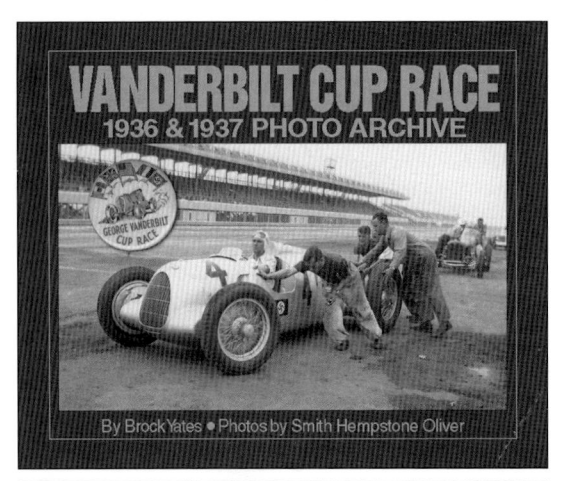

More great books from
Iconografix

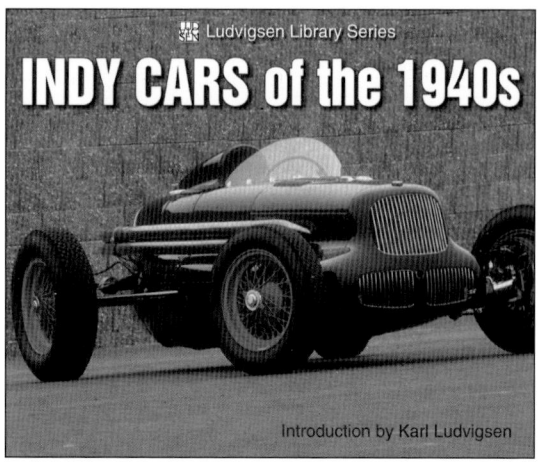

Vanderbilt Cup Race 1936 & 1937 Photo Archive ISBN 1-882256-66-2

Indy Cars of the 1940s Ludvigsen Library Series ISBN 1-58388-117-4

Indianapolis Racing Cars of Frank Kurtis 1941-1963 Photo Archive
ISBN 1-58388-026-7

Indy Cars of the 1950s Ludvigsen Library Series ISBN 1-58388-018-6

Lost Race Tracks Treasures of Automobile Racing ISBN 1-58388-084-4

Indy Cars of the 1970s Ludvigsen Library Series ISBN 1-58388-098-4

Indy Cars of the 1960s Ludvigsen Library Series ISBN 1-58388-052-6

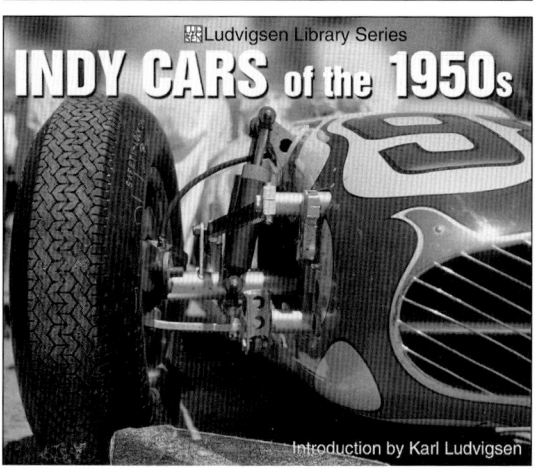

Iconografix, Inc.
P.O. Box 446, Dept BK,
Hudson, WI 54016
For a free catalog call: 1-800-289-3504

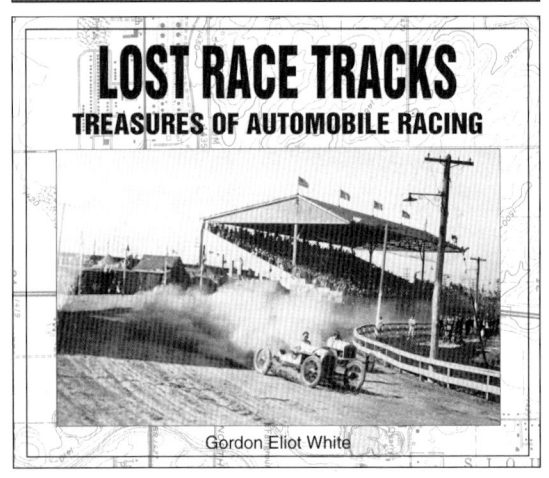

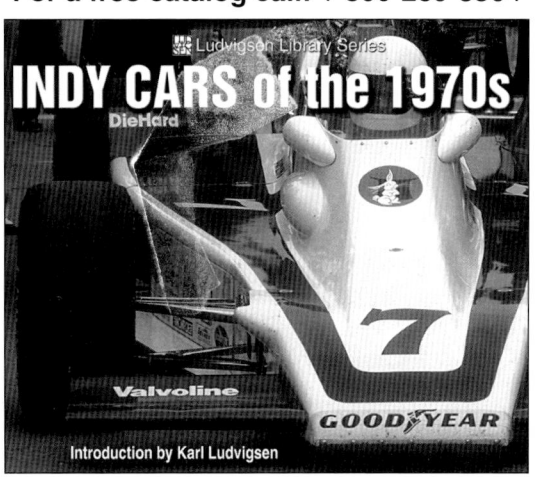

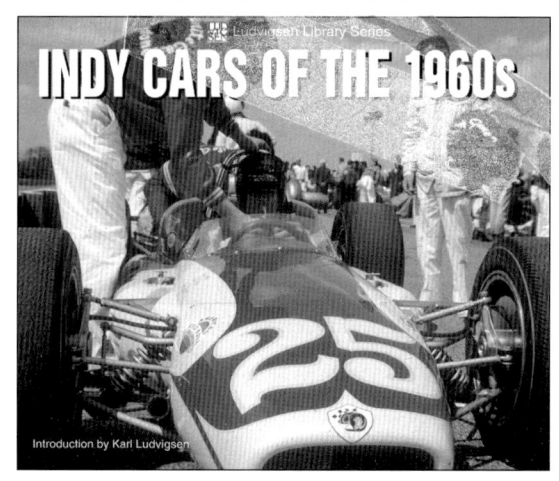